Sustainability in Europe

Radka MacGregor Pelikánová

Sustainability in Europe

Roots and Evolution of the Current Legal Regime

Springer

Radka MacGregor Pelikánová
Metropolitan University Prague
Prague, Czech Republic

ISBN 978-3-031-74688-8 ISBN 978-3-031-74689-5 (eBook)
https://doi.org/10.1007/978-3-031-74689-5

This Springer imprint is published by the registered company Springer Nature Switzerland AG
The registered company address is: Gewerbestrasse 11, 6330 Cham, Switzerland

Acknowledgments

This monograph is the result of Metropolitan University Prague research project no. 110-2 “Economic studies, International Business and Financial Management” (2024) based on a grant from the Institutional Fund for the Long-term Strategic Development of Research Organizations.

This monograph was completed thanks to the long-lasting and patient support of my excellent mentors, not only in the field of economics, law and intellectual property, JUDr. Karel Čermák and JUDr. Václav Pelikán, and of course my husband, Robert K. MacGregor, MBA. I am tremendously grateful to them and gladly dedicate this monograph to them.

In addition, I would like to express my gratitude for an over one decade long general academic support provided to me by the Metropolitan University of Prague and for a specific assistance by Mag. Roman Gassenbauer and Dr. Susanne Casagrande from Johannes Kepler University in Linz, and Dr. Sergio Ginebri, Dr. Roberto Baretta and Dr. Fiona Elisabeth MacMillan from University Roma Tre and Prof. Michael P. Seng from University of Illinois, Chicago, School of Law for providing me with great conditions to write this monograph as well as with information sources and feedback. I also wish to thank the editors at Springer for their commitment and advice during the writing of this book, in particular to Laura Hofmann.

Legal Documents

ISO Standards

ISO 14000 Environmental Management https://www.iso.org/standards/popular/iso-14000-family

ISO 14000.1 Environment Management Systems https://www.iso.org/standard/60857.html

ISO 26000 Social Responsibility – over 400 recommendations https://iso26000.info/

UN and OECD instruments

Charter of the United Nations signed on 26 June 1945, in San Francisco, at the conclusion of the United Nations Conference on International Organization ("UN Charter")

UN General Assembly Resolution A/RES/217(III) The Universal Declaration of Human Rights from 10 December 1948 ("UDHR")

UN Annex to document A/42/427 Report of the World Commission on Environment and Development: Our Common Future from 4 August 1987 ("Brundtland Report")

UN Conference on Environment and Development in Rio de Janeiro Declaration Agenda 21 from 3 to 14 June 1992 ("Agenda 21")

The Global Compact: Who Cares Wins. Connecting Financial Markets to a Changing World from 24 June 2004 ("Report Who Cares Wins")

SG/2111 ECO/106 Six Principles for Responsible Investment from 27 April 2006 launched by UN backed by world's largest investors and endorsed at New York Stock Exchange ("Six Principles")

International network of financial institutions endorsing and supporting the implementation of Six Principles supported by the UN ("UN PRI")

UN General Assembly Resolution A/RES/71/1 Transforming our world: the 2030 Agenda for Sustainable Development from 25 September 2015—People, Planet, Prosperity, Peace, and Partnership ("Agenda 2030")

Paris Agreement under the United Nations Framework Convention on Climate Change from 12 December 2015, Chapter XXVII Environment, 7 d., UN Treaty Series, vol. 3156, p.7 ("Paris Climate Agreement ")

OECD (2023), OECD Guidelines for Multinational Enterprises on Responsible Business Conduct, 8 June, 2023, OECD Publishing, Paris. https://doi.org/10.1787/81f92357-en ("OECD Guidelines 2023")

EU Legislative Instruments

Treaty on European Union, signed at Maastricht on 7 February 1992 (92/C 191/01) ("Treaty of Maastricht")

Treaty of Amsterdam amending the Treaty on European Union, the Treaties establishing the European Communities and certain related acts, signed at Amsterdam on 2 October 1997 (97/C 340/01-05) ("Treaty of Amsterdam")

Treaty of Lisbon amending the Treaty on European Union and the Treaty establishing the European Community, signed at Lisbon on 13 December 2007 (2007/C 306/01) ("Treaty of Lisbon")

Directive 2013/34/EU of the European Parliament and of the Council of 26 June 2013 on the annual financial statements, consolidated financial statements and related reports of certain types of undertakings, amending Directive 2006/43/EC of the European Parliament and of the Council and repealing Council Directives 78/660/EEC and 83/349/EEC ("Accounting Directive")

Directive 2014/95/EU of the European Parliament and of the Council of 22 October 2014 amending Directive 2013/34/EU as regards disclosure of non-financial and diversity information by certain large undertakings and groups ("NFRD")

Treaty on EU (C 202/1, OJ C 202 7.6.2016, p. 13)—consolidated in 2020 ("TEU")

Treaty on the Functioning of the EU (C 202/01, OJ C 202 7.6.2016, p. 47)—consolidated in 2020 ("TFEU")

Charter of Fundamental Rights of the European Union (2016/C 202/02, OJ C 202, 7.6.2016, p. 389-405) ("Charter")

Regulation (EU) 2019/2088 of the European Parliament and of the Council of 27 November 2019 on sustainability-related disclosures in the financial services sector ("SFDR")

Regulation (EU) 2020/852 of the European Parliament and of the Council of 18 June 2020 on the establishment of a framework to facilitate sustainable investment, and amending Regulation (EU) 2019/2088 ("Taxonomy Regulation")

Regulation (EU) 2021/1119 of the European Parliament and of the Council of 30 June 2021 establishing the framework for achieving climate neutrality and amending Regulations (EC) No 401/2009 and (EU) 2018/1999 ("European Climate Law")

Directive (EU) 2022/2464 of the European Parliament and of the Council of 14 December 2022 amending Regulation (EU) No 537/2014, Directive 2004/109/EC, Directive 2006/43/EC and Directive 2013/34/EU, as regards corporate sustainability reporting ("CSRD")

Commission Delegated Regulation (EU) 2023/2772 of 31 July 2023 supplementing Directive 2013/34/EU of the European Parliament and of the Council as regards sustainability reporting standards ("ESRS Regulation")

Directive (EU) 2024/1760 of the European Parliament and of the Council of 13 June 2024 on corporate sustainability due diligence and amending Directive (EU) 2019/1937 and Regulation (EU) 2023/2859 ("CSDDD" or "CS3D")

EU Case Law

C-62/88 Greece v. Council (Chernobyl I)

C-281/01 Commission v. Council (Energy-efficient labelling)

C-378/08 Raffineire Mediterranée Raffinerie Mediterranee (ERG) SpA (Augusta Sicily decontamination)

C-343/09 Afton Chemical Limited v Secretary of State for Transport (Contains metallic additives)

C-2/15 Opinion of the Court on the Free Trade Agreement between the EU and the Republic of Singapore (criterion of the sustainable development)

C-148/23 Gestore dei Servizi Energetici SpA – GSE v Erg Eolica (Renewable Energy Roadmap)

T-74/00 Artegodan v Commission (Tenuate Retard)

EU Policy Instruments

Lisbon European Council on 23 and 24 March 2000 Presidency Conclusions ("Lisbon Strategy")

COM(2001) 366 final Green Paper Promoting a European framework for Cor porate Social Responsibility from 18 July 2001 ("Green Paper for CSR")

COM(2002) 347 Final Communication from the Commission concerning Corporate Social Responsibility: A business contribution to Sustainable Development from 2 July 2002 ("Communication about CSR")

COM(2006) 136 Final Communication Implementing the Partnership for Growth And Jobs: Making Europe a Pole of Excellence on Corporate Social Responsibility from 22 March 2006

COM(2010) 2020 Final Communication from the Commission: EUROPE 2020 A strategy for smart, sustainable and inclusive growth from 3 March 2010 ("Europe 2020")

COM(2016) 739 Final Communication from the Commission to the European Parliament, the Council, the European Economic and Social Committee and the Committee of the Regions: Next steps for a sustainable European future—European action for sustainability. SWD(2016) 390 final from 22 November 2016 ("European Action for Sustainability" or "Next steps")

COM(2019) 22 final Reflection Paper Towards a Sustainable Europe by 2023 from 30 January 2019 ("Reflection Paper")

Von der Leyen U (2019) Political Guidelines for the next European Commission 2019-2024: A Union that strives for more. My agenda for Europe. Presented to the Members of European Parliament on 16 July 2019. Available from: https://www.europarl.europa.eu/resources/library/media/20190716RES57231/20190716RES57231.pdf ("Political Guidelines")

COM(2019) 640 Final Communication from the Commission to the European Parliament, the European Council, the Council, the European Economic and Social Committee and the Committee of the Regions: The European Green Deal from 11 December 2019 ("European Green Deal")

COM(2020) 98 Final Communication from the Commission to the European Parliament, the Council, the European Economic and Social Committee and the Committee of the Regions: A New Circular Economy Action Plan—For a cleaner and more competitive Europe from 11 March 2020 ("CEAP")

Von der Leyen U (2024) Statement at the European Parliament Plenary by President Ursula von der Leyen, candidate for a second mandate 2024-2029. Our first priority will be prosperity and competitiveness from 18 July 2024. Available from: https://neighbourhood-enlargement.ec.europa.eu/news/statement-european-parliament-plenary-president-ursula-von-der-leyen-candidate-second-mandate-2024-2024-07-18_en ("Statement for 2024-2029")

Von der Leyen U (2024) Political Guidelines for the next European Commission 2024-2029: Europe's Choice, 18 July 2024. Available from: https://commission.europa.eu/document/e6cd4328-673c-4e7a-8683-f63ffb2cf648_en ("New Political Guidelines")

Babylonian Law

Code of Hammurabi, 1755 - 1750 B.C., a collection of 282 rules

English (UK) Law

An Acte againste suche persones as doo make Bankrupte 34 & 35, 8th Parliament of Henry VIII, c. 4 of 1542 ("Statute of Bankrupts of 1542")

Statutory Instrument 2008 No. 1276 Business protection from Misleading Marketing Regulations 2008

Modern Slavery Act 2015 c.30 from 26 March 2015

French Law

Codex Vigilanus aka *Codex Albeldensis* of 976

L´ordonnance définissant le rôle des maîtres des Eaux et Forêts, 1291

L´ordonnance de Brunoy, 1346

Grande Ordonnance forestiere sur le fait des Eaux et Forêts, 1669

German Law

Dampfkessel-Gesetzgebung in Preußen von 1831

Preußische Gewerbeordnung von 1845

Reichsgewerbeordnung von 1871

Gesetz, betreffend die Krankenversicherung der Arbeiter (KVG) vom 15.Juni 1883 - Deutsches Reichsgesetzblatt Band 1883, Nr. 9, Seite 73–104 ("German Health Insurance Act 1883")

Unfallversicherungsgesetz vom 6.Juli 1884 - Deutsches Reichsgesetzblatt Band 1884, Nr. 19, Seite 69 ("German Accident Insurance Act 1884")

Einschränkung der Frauenarbeit und Kinderarbeit Gesetz von 17.Juni 1887 Deutsches Reichsgesetzblatt Band 1887, Nr. 88, Seite 913 ("Working Hours Limitation Act 1887")

Gesetz betreffend die Invaliditäts- und Altersversicherung vom 22.Juni 1889 - Deutsches Reichsgesetzblatt Band 1889, Nr. 13, Seite 97 ("German Old Age and Invalidity Insurance Act 1889")

Roman Law

Lex Duodecim Tabularum – Duodecim Tabulae (Law of the Twelve Tables), 449 BC

Corpus Iuris Civilis (Codex, Digest or Pandects, Institutes/Institutiones+ Novellae Constitutiones), 529-534

USA Law

Act to protect trade and commerce against unlawful restraints and monopolies, 15 U.S.C. §§ 1–7, from 2nd July 1890 ("Sherman Antitrust Act 1890")

An Act to repeal timber-culture laws, and for other purposes, 16 U.S.C. ch. 2 from 3rd March 1891 ("Forest Reserve Act of 1891")

Antitrust Act, 15 U.S.C. §§ 12–27, 29 U.S.C. §§ 52–53, from 15th October 1911 ("Clayton Act 1914")

Act to create a Federal Trade Commission, to define its powers and duties, and for other purposes, 38 Stat. 717, from 26th September 1914 ("Federal Trade Commission Act 1914")

Fair Housing Act, 42 U.S.C. 3601 et seq., from 11th April 1968 ("Fair Housing Act 1968")

Community Reinvestment Act, 12 U.S.C. § 2901 et seq. from 12th October 1977 ("Community Reinvestment Act 1977")

Contents

About the Author

Radka MacGregor Pelikánová Ph.D., LL.M., MBA (ORCID 0000-0001-9628-7146) is a senior lecturer at Metropolitan University Prague (MUP) and a research lead at the University of New York in Prague (UNYP) with an expertise in law, economics, management, business and intellectual property. She completed the study of law at Charles University School of Law in Prague (Mgr., Ph.D., JUDr.), Université Paris XIII—Villetaneuse in Paris (licence) and Wayne State University in Detroit (LL.M.). In addition, she completed the study of accounting at Washtenaw Community College (Accounting certificate—CTACC) and of economics at the National American University in cooperation with the UNYP (MBA). For over two decades, she has been a member of the Czech Bar Association and the Michigan Bar Association. She has a strong and long-lasting experience providing clients with advice regarding intellectual property, economics and law. During the last fifteen years, she has been focusing on academic and publication activities, especially regarding Intellectual Property (IP), EU law, Competition law, Business law, Strategic Management and Knowledge Management. In addition, she is a legal and strategic management advisor at the Ministry of Agriculture of the Czech Republic. She has published over twenty high quality multidisciplinary articles about sustainability, corporate social responsibility (CSR) and creating shared values (CSV) while focusing on the evolution of the interaction of their economic and law aspects in Europe.

Abbreviations

Accounting Directive	Directive 2013/34/EU of the European Parliament and of the Council of 26 June 2013 on the annual financial statements, consolidated financial statements and related reports of certain types of undertakings, amending Directive 2006/43/EC of the European Parliament and of the Council and repealing Council Directives 78/660/EEC and 83/349/EEC ("Accounting Directive")
Agenda 21	UN Conference on Environment and Development in Rio de Janeiro Declaration Agenda 21 from 3 to 14 June 1992 ("Agenda 21")
Agenda 2030	UN General Assembly Resolution A/RES/71/1 Transforming our world: The 2030 Agenda for Sustainable Development from 25 September 2015—People, Planet, Prosperity, Peace, and Partnership ("Agenda 2030")
Brundtland Report	UN Annex to document A/42/427 Report of the World Commission on Environment and Development: Our Common Future from 4 August 1987 ("Brundtland Report")
CEAP	COM(2020) 98 Final Communication from the Commission to the European Parliament, the Council, the European Economic and Social Committee and the Committee of the Regions: A New Circular Economy Action Plan—For a cleaner and more competitive Europe from 11 March 2020 ("CEAP")

Charter	Charter of Fundamental Rights of the European Union (2016/C 202/02, OJ C 202, 7.6.2016, p. 389-405) ("Charter")
CJ EU	Court of Justice of EU
Clean Industrial Deal	Clean Industrial Deal (new) announced by the Statement for 2024–2029
Communication about CSR	COM(2002) 347 Final Communication from the Commission concerning Corporate Social Responsibility: A business contribution to Sustainable Development from 2 July 2002 ("Communication about CSR")
CS3D	Directive (EU) 2024/1760 of the European Parliament and of the Council of 13 June 2024 on corporate sustainability due diligence and amending Directive (EU) 2019/1937 and Regulation (EU) 2023/2859 ("CSDDD" or "CS3D")
CSI	Corporate Social Irresponsibility
CSDDD	Directive (EU) 2024/1760 of the European Parliament and of the Council of 13 June 2024 on corporate sustainability due diligence and amending Directive (EU) 2019/1937 and Regulation (EU) 2023/2859 ("CSDDD" or "CS3D")
CSR	Corporate Social Responsibility
CSRD	Corporate Social Reporting Directive (EU) 2022/2464 of the European Parliament and of the Council of 14 December 2022 amending Regulation (EU) No 537/2014, Directive 2004/109/EC, Directive 2006/43/EC and Directive 2013/34/EU, as regards corporate sustainability reporting ("CSRD")
CSV	Creating Shared Value
EGD	COM(2019) 640 Final Communication from the Commission to the European Parliament, the European Council, the Council, the European Economic and Social Committee and the Committee of the Regions: The European Green Deal from 11 December 2019 ("European Green Deal")
ESG	Environment Social Governance
ESRS	European sustainability reporting standards included in Annex I of the ESRS Regulations

ESRS Regulation	Commission Delegated Regulation (EU) 2023/2772 of 31 July 2023 supplementing Directive 2013/34/EU of the European Parliament and of the Council as regards sustainability reporting standards
Europe 2020	COM(2010) 2020 Final Communication from the Commission: EUROPE 2020 A strategy for smart, sustainable and inclusive growth from 3 March 2010 ("Europe 2020")
European Action for Sustainability	COM(2016) 739 Final Communication from the Commission to the European Parliament, the Council, the European Economic and Social Committee and the Committee of the Regions: Next steps for a sustainable European future—European action for sustainability. SWD(2016) 390 final from 22 November 2016 ("European Action for Sustainability" or "Next steps")
European Climate Law	Regulation (EU) 2021/1119 of the European Parliament and of the Council of 30 June 2021 establishing the framework for achieving climate neutrality and amending Regulations (EC) No 401/2009 and (EU) 2018/1999 ("European Climate Law")
Green Paper for CSR	COM(2001) 366 final Green Paper Promoting a European framework for Corporate Social Responsibility from 18 July 2001 ("Green Paper for CSR")
GRI	Global Reporting Initiative—a non-governmental organization founded in 1997 in Boston
GRI Standards	Standards issued by GRI
IASB	International Accounting Standards Board
IFRS Foundation	International Financial Reporting Standards Foundation which establishes and develops globally accepted accounting and sustainability disclosure standards while using IASB and ISSB
ISO	International Organization for Standardization
ISSB	International Sustainability Standards Board
Lisbon Strategy	Lisbon European Council on 23 and 24 March 2000 Presidency Conclusions ("Lisbon Strategy")

New Political Guidelines	Von der Leyen U (2024) Political Guidelines for the next European Commission 2024-2029: Europe´s Choice, 18 July 2024. Available from: https://commission.europa.eu/document/e6cd4328-673c-4e7a-8683-f63ffb2cf648_en ("New Political Guidelines")
Next steps	COM(2016) 739 Final Communication from the Commission to the European Parliament, the Council, the European Economic and Social Committee and the Committee of the Regions: Next steps for a sustainable European future—European action for sustainability. SWD(2016) 390 final from 22 November 2016 ("European Action for Sustainability" or "Next steps")
NFRD	Directive 2014/95/EU of the European Parliament and of the Council of 22 October 2014 amending Directive 2013/34/EU as regards the disclosure of non-financial and diversity information by certain large undertakings and groups Non-Financial Reporting Directive ("NFRD")
OECD	Organisation for Economic Co-operation and Development which replaced in 1961 OEEC ("OECD")
OECD Guidelines 2023	OECD (2023) Guidelines for Multinational Enterprises on Responsible Business Conduct, OECD Publishing, Paris ("OECD Guidelines 2023")
OEEC	Organisation for European Economic Co-operation formed in 1948 and in 1961 replaced by OECD ("OEEC")
Paris Climate Agreement	Paris Agreement under the United Nations Framework Convention on Climate Change from 12 December 2015, Chapter XXVII Environment, 7 d., UN Treaty Series, vol. 3156, p.7 ("Paris Climate Agreement")
Political Guidelines	Von der Leyen U (2019) Political Guidelines for the next European Commission 2019-2024: A Union that strives for more. My agenda for Europe. Presented to the Members of European Parliament on 16 July 2019. Available from: https://www.europarl.europa.eu/resources/library/media/20190716RES57231/20190716RES57231.pdf ("Political Guidelines")

Reflection Paper	COM(2019) 22 final Reflection Paper Towards a Sustainable Europe by 2023 from 30 January 2019 ("Reflection Paper")
Report Who Cares Wins	Global Compact: Who Cares Wins. Connecting Financial Markets to a Changing World from 24 June 2004
SASB	Sustainability Accounting Standards Board
SDG	Sustainable Development Goal
SFDR	Regulation (EU) 2019/2088 of the European Parliament and of the Council of 27 November 2019 on sustainability-related disclosures in the financial services sector ("SFDR")
Six Principles	SG/2111 ECO/106 Six Principles for Responsible Investment from 27 April 2006—launched by the UN, backed by world's largest investors and endorsed by the New York Stock Exchange
Statement for 2024-2029	Von der Leyen U (2024) Statement at the European Parliament Plenary by President Ursula von der Leyen, candidate for a second mandate 2024–2029. Our first priority will be prosperity and competitiveness from 18 July 2024. Available from: https://neighbourhood-enlargement.ec.europa.eu/news/statement-european-parliament-plenary-president-ursula-von-der-leyen-candidate-second-mandate-2024-2024-07-18_en ("Statement for 2024-2029")
Taxonomy Regulation	Regulation (EU) 2020/852 of the European Parliament and of the Council of 18 June 2020 on the establishment of a framework to facilitate sustainable investment, and amending Regulation (EU) 2019/2088 ("Taxonomy Regulation")
TEU	Treaty on EU (C 202/1, OJ C 202 7.6.2016, p. 13)—consolidated in 2020 ("TEU")
TFEU	Treaty on the Functioning of the EU (C 202/01, OJ C 202 7.6.2016, p. 47)— consolidated in 2020 ("TFEU")
Treaty of Maastricht	Treaty on European Union, signed at Maastricht on 7 February 1992 (92/C 191/01) ("Treaty of Maastricht")

Treaty of Amsterdam	Treaty of Amsterdam amending the Treaty on European Union, the Treaties establishing the European Communities and certain related acts, signed at Amsterdam on 2 October 1997 (97/C 340/01-05) ("Treaty of Amsterdam")
Treaty of Lisbon	Treaty of Lisbon amending the Treaty on European Union and the Treaty establishing the European Community, signed at Lisbon on 13 December 2007 (2007/C 306/01) ("Treaty of Lisbon")
UDHR	UN General Assembly Resolution A/RES/217 (III) The Universal Declaration of Human Rights from 10 December 1948 ("UDHR")
UN	United Nations
UN Charter	Charter of the United Nations signed on 26 June 1945, in San Francisco, at the conclusion of the United Nations Conference on International Organization ("UN Charter")
UN PRI	International network of financial institutions endorsing and supporting the implementation of Six Principles supported by the UN
WCED	World Commission on Environment and Development

Chapter 1
Introduction

Well, this introduction starts by discussing whether there is a general and universal responsibility for sustainability (1.) and it moves to address the international law crusade by the UN against depletion of resources and poverty (2.) and the private law crusade via CSR launched by American academics and businessmen (3.). Then, it deals with European style of sustainability (4.) and considers the future (5). Such an introduction gradually leads to the plan of this monograph via ten chapters to answer both questions, i.e. about the meaning and legal regime of sustainability in Europe. Pursuant to such a plan, after the Introduction (1st Chapter), sources and methods are to identified (2nd Chapter.), the terminology and scientific background set (3rd Chapter.) and its origins presented (4th Chapter). There follows a multi-disciplinarity presentation of the evolution of sustainability (5th Chapter) and its modern history (6th Chapter), including dynamic aspects of sustainable development (7th Chapter), a multi-stakeholder model and CSR (8th Chapter.) and the enforcement via the ESG (9th Chapter.). This plethora of information offers answers to both questions with pioneering conclusions (10th Chapter.).

"*Tempora mutantur et nos mutamur in iis*." Pursuant to this iconic and heavily cited Latin proverb the times change and we change with them and thus we are undergoing the Aristotelian ongoing motion and change while maintaining the substance. Nature is not something external to us[1] and the evolution of humans has been taking place within multiple webs of scientific, technological and other connections, influenced by culture and creativity,[2] within and between intersecting spheres.[3] Since humans have developed their genetic predisposition to quickly recognize environments rich in resources and exploit them in an intense and even exhausting and selfish manner,[4] evolution is marked by numerous positive and

[1] Dodman et al. (2020), pp. 3–10.

[2] Fogarty et al. (2015), pp. 736–754.

[3] Dodman et al. (2020), pp. 3–10.

[4] Barbiero (2021), pp. 43–78.

R. MacGregor Pelikánová, *Sustainability in Europe*,
https://doi.org/10.1007/978-3-031-74689-5_1

negative consequences of such behaviour patterns.[5] Indeed, humans have developed their inclination to transform the environment permanently and irreversibly,[6] their capacity to build and share knowledge,[7] their creativity to manage finance, develop the economy and resolve problems while using technology[8] and their use of tangible resources, leading to the exhaustion and collapse of our civilization. As a matter of fact, arguably the most serious challenge, of right now, is the climate change caused by our (human) actions.[9]

This leads to the concern whether we have a sustainable Europe, or more narrowly, whether there is sustainability in Europe and if yes, then what kind of sustainability and with what regime. Addressing such a complex and challenging concern is a Sisyphean task calling for a systematic step-by-step approach. Therefore, the research aim of this monograph is to explore the roots and evolution of sustainability in order to contextually figure out the development of its meaning and regime in Europe. This aim is translated into two mutually related goals, or, to put it better, questions: (i) what is the foundation, development and meaning of the concept of sustainability in Europe and (ii) what kind of regime and duties has it involved in Europe? Namely, once we understand properly the concept of sustainability, we can create, interpret and apply legitimately, effectively and efficiently its legal regime, and, in particular, the duties generated by the set obligations. Since sustainability is about the balanced ability to continue based on a pragmatic respect for the future, it is rather an ephemeral idea with an inherent subjective meaning. At the same time, as we currently see, the concept of sustainability is critical for the continuation of our civilization and so it needs to be recognized as a legal concept with a proper regime, i.e. as leading to a title generating obligations and this means both legally enforceable rights and duties.

So, do we have a general and universal responsibility for sustainability (Sect. 1.1)? Well, the UN fights against the very bad depletion of resources, against poverty and injustice (Sect. 1.2), while the Americans are trying to address these burning issues by various methods and instruments involving the large public, in particular businesses (Sect. 1.3). Europe has its own particular way, leading to the European style of sustainability (Sect. 1.4). The most important is the future - the future of sustainability, Europe, our civilization and perhaps the entire humanity and our planet. Quo Vadis? (Sect. 1.5)

[5] MacGregor Pelikánová et al. (2024b), pp. 1–29.

[6] Barbiero (2021), pp. 43–78.

[7] Dodman et al. (2020), pp. 3–10.

[8] MacGregor Pelikánová and Sani (2023), pp. 39–60.

[9] Jones et al. (2023), p. 4.

1.1 General and Universal Responsibility for Sustainability?

Indeed, we, humans, are subjects with rights and duties and so we have to face up to the obligation to protect the objects and subject matters so we as subjects can continue, ideally forever and this perpetuity might be symbolically visualized as the eternal circle. This is the perception heralded by European civilization which is strongly shaped by and has foundations in Christianity[10] and is intimately linked to the concept of responsibility[11] and the concept of sustainability.[12]

Propositions about individual and collective responsibility for actions and omissions have impacted society from antiquity to today and they were often paralleled with demands to preserve the continuous existence of human kind, which reflects predominantly the continental law tradition. The setting of sustainability, as well as responsibility, has always mirrored value judgments about justice and is suitable for functionalist, analytical, contextual and historical comparisons.[13] For sustainability, it was and is about justice in distributing and using resources. For responsibility, it was and is about justifying one's own behaviour and bearing the consequences—either informally towards oneself (accountability), or ethically towards a determined group (responsibility) or legally towards the entire society (liability). Therefore, the intersection of moralizing ethics, restrictive and enforceable laws and pragmatic economics was, and remains, obvious and makes sustainability a relevant target for studies by various branches of social science.

There is an inevitable tension between the legal positivism of Hart's will,[14] the compromises of Hobbes' social contract, the utilitarian concession of Bentham's pragmatism, the deontological principles of Kant's judgment[15] and the universalism of Aristotle's law of nature and subjective ethical features reflecting one's own personal morals. This makes sustainability and the attitude to it ephemeral and strongly contextually and historically dependent. Plainly, sustainability is obscure and its interpretation extremely diversified, often reduced to empirical observations,[16] but at the same time omnipresent and inevitable—for thousands of years.

Indeed, despotic monarchs, from hundreds to a thousand years past, as well as decision makers today, realized that they needed to take steps to preserve the ongoing existence of their realms. This burning and complex task has been approached from diverse angles and through different strategies while involving different stakeholders and setting varying goals.

---

[10] Stark (2014), pp. 2–5.

[11] MacGregor Pelikánová (2017), pp. 81–98.

[12] Stark (2009), pp. 8–9.

[13] Tóthová (2023), pp. 78–88Ind.

[14] Stamper (1991), pp. 219–44.

[15] Balcerzak and MacGregor Pelikánová (2020), p. 95.

[16] White (2013), pp. 213–217.

At the very beginning, it was a pragmatic local need to manage natural resources to allow for an ongoing agricultural production and this accountability for such a need was borne typically by the leader while other social and environmental concerns played a small role, if any. The Mesopotamian and Egyptian flooding, cultivation and storage management by such leaders is a perfect example. The accountability of these leaders often had a religious dimension, i.e. the leader had to account for his management of those resources entrusted to him by God to himself, while the clan or tribe practiced a collective accountability with the enhanced reciprocity, see "If a man knocks out the teeth of his equal, his teeth shall be knocked out," aka "a tooth for a tooth." Christianity brought about the individual responsibility and extension of concerns, i.e. the management, maintenance and distribution of resources was to be done while considering more than just the locals and while not only keeping accounts, but, as well, being responsible for them. The cosmopolitan Roman Empire and the Roman law, especially during the age of the Emperors, made a clear move towards regional and super-regional concerns with individual responsibility often becoming even enforceable by law, i.e. liability.

The trend towards broadening the pool of resources and subjects and towards extending the responsibility was revived during the Middle-Ages and the Industrial Revolution shifted the focus pendulum from agriculture to industrialization. It then started to become clear that, not only is it necessary to keep renewing resources (ongoing economics) but as well to protect them (ongoing environment) by achieving social peace (ongoing social).

At the same time, the involvement of all members of society became a necessity, see the United Nations ("UN") with the Universal Declaration of Human Rights ("UDHR")[17] from 1948 stating "Everyone has duties to the community in which alone the free and full development of his personality is possible." (Art. 29 UDHR) and the expansion of activities of the Organisation for European Economic Co-operation ("OEEC"), in particular after its replacement by the Organisation for Economic Co-operation and Development ("OECD").[18] A setting, such as this, was getting open to the deeper consideration of other factors, such as resource management and justice in distribution, at least on the international (law) level.

[17] UN General Assembly Resolution A/RES/217(III) The Universal Declaration of Human Rights from 10 December 1948 ("UDHR").

[18] Carolan (2008), pp. 51–65.

1.2 UN Against Bad Depletion and Poverty "Which Is an Evil in Itself"

A foundation stone for the modern concept of sustainability was placed in 1972 via the famous book by Meadows stating "If the present growth trends in world population, industrialization, pollution, food production, and resource depletion continue unchanged the limits to growth on this planet will be reached sometime within the next one hundred years."[19] Consequently, the modern concept of sustainability fully emerged in 1987 on the international public macro-level under the auspices of the UN through the famous report "Our Common Future" prepared by the World Commission on Environment and Development ("WCED") of Dr. Gor Harlem Brundtland ("Brundtland Report")[20] and got cemented in 1992 by Agenda 21 ("Agenda 21").[21]

The motto of the Brundtland Report "poverty is an evil in itself" was made regarding the entire world population in Part I Chapter 1 Point 8 of the Brundtland Report. Further, the definition of sustainable development included in Part I Chapter 3 Point 27 of the Brundtland Report is perhaps one of the most cited definitions in the EU and perhaps even globally—"Humanity has the ability to make development sustainable to ensure that it meets the needs of the present without compromising the ability of future generations to meet their own needs. The concept of sustainable development does imply limits - not absolute limits but limitations imposed by the present state of technology and social organization on environmental resources and by the ability of the biosphere to absorb the effects of human activities."

The Brundtland report brought, or at least launched, three conceptual innovations. Firstly, it pointed to the capacity to evolve eternally towards reaching an increased efficiency resource usage for the benefit of humanity and other species and thus added the feature of a massive efficiency (not effectiveness). Consequently, sustainability was not only about maintaining local resources for the local population, but about providing all people globally wide with a basic quality of life while protecting the ecosystem. Sustainability turned out to be about human progress at a global level for the long term, i.e. a durable extensiveness worldwide.[22] Secondly, the iconic three pillars structure, i.e. environmental (planet), social (people) and economic (profit) dimensions,[23] was implied by the Brundtland Report and more explicitly

[19] Meadows et al. (1972).

[20] UN Annex to document A/42/427 Report of the World Commission on Environment and Development: Our Common Future from 4 August 1987 ("Brundtland Report").

[21] UN Conference on Environment and Development in Rio de Janeiro Declaration Agenda 21 from 3 to 14 June 1992 ("Agenda 21").

[22] Cristian et al. (2015), pp. 806–811.

[23] Purvis et al. (2019), pp. 681–695.

expressed by Agenda 21.[24] Thirdly, the cross-generational feature of sustainability making it a concept bridging generations and establishing a solidarity between past, current and future generations of human beings was cemented.

The evolution of the modern concept of sustainability on the international level reached its latest milestone in 2015 with the UN Agenda for Sustainable development 2030 ("Agenda 2030"),[25] which brought forth the famous 17 Sustainable Development Goals ("SDGs") and 169 targets. The EU and many other countries promptly endorsed these trends and enthusiastically followed Agenda 2030 and embraced SDGs. However, this interest and commitment was neither shared nor trusted by all[26] and many international law subjects or international leaders have repeatedly emphasized their mistrust and/or rejection and/or ignorance of sustainability and sustainability issues.

1.3 US-American CSR, CSV, CSI and ESG

The feasibility and success of the modern concept of sustainability cannot be achieved only via international public macro-endeavors and, already prior to the Brundtland Report, there began a discussion about the need for a multi-stakeholder model and active partnership with not only international law subjects but as well national subjects. The voices calling for the involvement and responsibility of large businesses, see CSR, which originated during the first half of the twentieth century in the USA, were growing progressively stronger and louder and more global.

After addressing Codes of Ethics[27] and similar instruments, the focus in the US was re-shaped by a set of highly influential academics and businessmen. In 1953 came Howard Rothmann Bowen's famous idea to hold (large) businesses not only accountable to themselves or perhaps responsible to their shareholders, but to society as a whole, see "The decisions and actions of the businessman have a direct bearing on the quality of our lives and personalities. His decisions affect not only himself, his stockholders, his immediate workers, or his customers – they affect the lives and fortunes of us all."[28] Consequently, sustainability started to be paralleled by the trade-off - corporate social responsibility ("CSR"), which was originally visualized by Archie B. Carroll via the famous pyramid diagram in 1991.[29] The advanced

[24] UN Conference on Environment and Development in Rio de Janeiro Declaration Agenda 21 from 3 to 14 June 1992 ("Agenda 21").

[25] UN General Assembly Resolution A/RES/71/1 Transforming our world: the 2030 Agenda for Sustainable Development from 25 September 2015 – People, Planet, Prosperity, Peace, and Partnership ("Agenda 2030").

[26] Harlow et al. (2013), pp. 270–280.

[27] Balcerzak and MacGregor Pelikánová (2020), p. 95.

[28] Bowen (1953), p. 3.

[29] Carroll (1999), pp. 268–295.

win-win form of CSR called Creating Shared Value ("CSV") was originally proposed by Michael Eugene Porter and Mark R. Kramer in 2006.[30]

Generally, CSR means that businesses have to meet society's expectations in their operations.[31] Pursuant to the underlying belief, CSR is the strategic process through which corporations can solve a social problem which is aligned to their value chain while concurrently pursuing economic profits. Arguably, CSR is about balancing the needs of (all) stakeholders during the given time, while the modern concept of sustainability is about balancing the needs of (all) current and future stakeholders and so overcoming the short-termism.[32] Consequently, the pool of CSR subjects and addressees keeps growing and expanding to include more businesses and more stakeholders, including consumers.[33] However, by no means should this be understood as the finding that the modern concept of sustainability is supported actively across the entire society without any reservations, see e.g. some famous CSR scandals demonstrating Corporate Social Irresponsibility ("CSI") and still inducing the support and solidarity by customers and investors.[34] Indeed, at least three key concepts for the management of a business and its competitiveness have co-existed: Friedman's traditional, aka conventional, shareholder model to maximize returns to owners (the ultimate responsibility of a business is to increase the profits of its shareholders),[35] Freeman's stakeholder model to take care of primary stakeholders while maintaining the business in good health,[36] and Carroll's CSR four responsibility layered pyramid model, to do good in society without necessarily aiming at profits.[37] The last mentioned CSR pyramid model, of which the most important responsibilities are allegedly the economic and ethical responsibilities,[38] has been recently further advanced towards the CSV.

The Agenda 2030 in 2015 and many related endeavours, including in the sphere of academia,[39] have been advocating for a multi-level partnership for sustainability and SDGs[40] and have definitely contributed to the move from CSR to CSV[41] and to their "measurable information" via Environment Social Governance ("ESG"). Since ESG criteria helps to identify material risks and growth opportunities and assists in

[30] Porter and Kramer (2011), pp. 62–77.

[31] Gössling and Vocht (2007), pp. 363–372.

[32] Bansal and DesJardine (2014), pp. 70–78.

[33] MacGregor Pelikánová (2024), pp. 218–235.

[34] Destri et al. (2024), pp. 1–26.

[35] Friedman (2007), pp. 173–178.

[36] Freeman (2017), pp. 449–465.

[37] Carroll (1999), pp. 268–295.

[38] Gössling and Vocht (2007), pp. 363–372.

[39] Van Tulder and Keen (2018), pp. 315–332.

[40] Van Tulder and Van Mil (2023), pp. 106–119.

[41] Kramer and Pfitzer (2016), pp. 80–89.

predicting both financial[42] and non-financial performance,[43] their over 30 years long presentation via scoring, including Dow Jones Sustainability indexes and London FTSE4-Good is highly important.[44] Pursuant to the underlying belief, there is a need for a non-hierarchical communication across the society[45] and that a viable (if not the only one) option is to tie both ends together, i.e. SDGs and social responsibility endeavours of businesses and individuals in an actionable manner and cross-sectorial partnership.

1.4 European Style of Sustainability

Historical, as well as modern, evolutions of sustainability have occurred predominantly in Europe and, unlike CSR with its common law roots, its roots and features are products of the continental law tradition based on the reason,[46] abstract legal doctrines[47] and Byzantine-Roman systematization,[48] and a partially paternalistic state approach.[49] The concept of sustainability developed along with the development of European states-to-be and nations-to-be, has acquired various national particularities and now the EU wants to consolidated it, or at least harmonized it if not standardized it.[50] Sustainability, the approach to it and its regime have undergone an evolution with European civilization, the importance of which has been increasingly recognized by societies and states in Europe. From a local pragmatic factor for an action done by a leader remaining above the law and not having any responsibility or liability, and being only accountable to the divinity, sustainability has been elevated to an ethical command generating a responsibility for regional elites and now is even turning into a legal command meaning the liability of an increasingly inclusive pool of subjects in the entire EU. The European style of sustainability aka Sustainability European Style is becoming matched by the Responsibility European Style of many Europeans.[51]

Indeed, if not before, then since the creation of the EU via the Treaty on European Union (92/C 191/01) ("Treaty of Maastricht"), sustainability started to make its way into the EU primary law, such as Treaty on EU ("TEU"),[52] Treaty on Functioning of

[42] Friede et al. (2015), pp. 210–233.

[43] Zatonatska et al. (2024), p. 2950.

[44] Gössling and Vocht (2007), pp. 363–372.

[45] MacGregor Pelikánová et al. (2024a), pp. 214–239.

[46] Stark (2014), pp. 39.

[47] MacGregor Pelikánová (2017), pp. 81–98.

[48] Brouwer (2018), pp. 1–11.

[49] Domański (2004), pp. 377–381.

[50] MacGregor Pelikánová (2024), pp. 218–235.

[51] De Schutter (2008), pp. 203–236.

[52] Treaty on EU (C 202/1, OJ C 202 7.6.2016, p. 13) – consolidated in 2020 ("TEU").

EU ("TFEU"),[53] and Charter of Fundamental Rights of the EU ("Charter"),[54] EU secondary law (see below) and even EU policies, such as Presidency Conclusions of Lisbon from 2000 ("Lisbon Strategy")[55] and the famous one decade strategy from 2010 ("Europe 2020").[56] Specifically, in 1992, the Treaty of Maastricht[57] set as one of the objectives "to promote economic and social progress which is balanced and sustainable" and in 1997, the Treaty of Amsterdam[58] set as one of the objectives "to achieve balanced and sustainable development." Consequently, for already three decades sustainability is explicitly included in the EU primary law as an objective to be followed and to be achieved, and this can be understood as an indirect legal command to sustainability. The direct command and reporting duty about sustainability both go back to the Non-Financial Reporting Directive ("NFRD")[59] in 2014 and keeps growing and expanding towards a positive substantive duty per se in lines with the Agenda 2030 and its 17 SDGs.

The EU, in particular the internal pro-integration tandem—the European Commission and the Court of Justice of EU ("CJ EU"), based on the above-mentioned law and policies, keeps moving further from proclamations, strategies and other instruments towards various general and special legal norms on sustainability and their implementation and application. Sustainability is becoming a genuine legal concept generating not only substantive obligations but, as well, having procedural and proceeding aspects, such as being an interpretation paradigm.

Indeed, a solid secondary EU law addresses sustainability indirectly, via a reporting duty as set by the Accounting Directive from 2013 ("Accounting Directive")[60] as updated via the already above mentioned NFRD[61] and Corporate

---

[53] Treaty on the Functioning of the EU (C 202/01, OJ C 202 7.6.2016, p. 47) – consolidated in 2020 ("TFEU").

[54] Charter of Fundamental Rights of the European Union (2016/C 202/02, OJ C 202, 7.6.2016, p. 389–405) ("Charter").

[55] Lisbon European Council on 23 and 24 March 2000 Presidency Conclusions ("Lisbon Strategy").

[56] COM(2010) 2020 Final Communication from the Commission: EUROPE 2020 A strategy for smart, sustainable and inclusive growth from 3 March 2010 ("Europe 2020").

[57] Treaty on European Union, signed at Maastricht on 7 February 1992 (92/C 191/01) ("Treaty of Maastricht").

[58] Treaty of Amsterdam amending the Treaty on European Union, the Treaties establishing the European Communities and certain related acts, signed at Amsterdam on 2 October 1997 (97/C 340/01-05) ("Treaty of Amsterdam").

[59] Directive 2014/95/EU of the European Parliament and of the Council of 22 October 2014 amending Directive 2013/34/EU as regards disclosure of non-financial and diversity information by certain large undertakings and groups ("NFRD").

[60] Directive 2013/34/EU of the European Parliament and of the Council of 26 June 2013 on the annual financial statements, consolidated financial statements and related reports of certain types of undertakings, amending Directive 2006/43/EC of the European Parliament and of the Council and repealing Council Directives 78/660/EEC and 83/349/EEC ("Accounting Directive").

[61] Directive 2014/95/EU of the European Parliament and of the Council of 22 October 2014 amending Directive 2013/34/EU as regards disclosure of non-financial and diversity information by certain large undertakings and groups ("NFRD").

Sustainability Reporting Directive ("CSRD"),[62] or even directly via setting the legal duty to sustainability, see the Sustainable Finance Disclosure Regulation ("SFDR").[63] Indeed, the drive to a more mandatory regime can be recently observed, see the Regulation setting the framework to facilitate sustainable investment ("Taxonomy Regulation")[64] and the Regulation about sustainability reporting standards ("ESRS Regulation").[65] Further, the drive to a more encompassing regime can be recently observed also, see the Regulation setting the framework for achieving climate neutrality ("European Climate law")[66] and supply chain reporting involving the Directive on corporate sustainability due diligence ("CSDDD" or "CS3D").[67]

The UN, with the Agenda 2030, along with the EU, kept demanding more actionability,[68] while sceptics showered them with criticism.[69] Since 2019, the academic and laic discussion seems even more agitated than before, partially due to the clash between the Political Guidelines of the first Commission of Ursula von der Leyen for 2019–2024 ("Political Guidelines"),[70] including the European Green Deal ("EGD"),[71] and the demand for freedom and justice by Europeans in the context of the Covid-19 pandemic[72] and the War in Ukraine[73] as well as other events dramatically changing the setting on the macro- as well as micro-level.[74]

---

[62] Directive (EU) 2022/2464 of the European Parliament and of the Council of 14 December 2022 amending Regulation (EU) No 537/2014, Directive 2004/109/EC, Directive 2006/43/EC and Directive 2013/34/EU, as regards corporate sustainability reporting ("CSRD").

[63] Regulation (EU) 2019/2088 of the European Parliament And of the Council of 27 November 2019 on sustainability-related disclosures in the financial services sector ("SFDR").

[64] Regulation (EU) 2020/852 of the European Parliament and of the Council of 18 June 2020 on the establishment of a framework to facilitate sustainable investment, and amending Regulation (EU) 2019/2088 ("Taxonomy Regulation").

[65] Commission Delegated Regulation (EU) 2023/2772 of 31 July 2023 supplementing Directive 2013/34/EU of the European Parliament and of the Council as regards sustainability reporting standards ("ESRS Regulation").

[66] Regulation (EU) 2021/1119 of the European Parliament and of the Council of 30 June 2021 establishing the framework for achieving climate neutrality and amending Regulations (EC) No 401/2009 and (EU) 2018/1999 ('European Climate Law').

[67] Directive (EU) 2024/1760 of the European Parliament and of the Council of 13 June 2024 on corporate sustainability due diligence and amending Directive (EU) 2019/1937 and Regulation (EU) 2023/2859 ("CSDDD" or "CS3D").

[68] Van Tulder et al. (2021), pp. 1–21.

[69] Anderson and Kosnik (2002), pp. 439–475.

[70] Von der Leyen U (2019) Political Guidelines for the next European Commission 2019–2024: A Union that strives for more. My agenda for Europe. Available from: https://www.europarl.europa.eu/resources/library/media/20190716RES57231/20190716RES57231.pdf ("Political Guidelines").

[71] COM/2019/640 final Communication from the Commission to the European Parliament, the European Council, the Council, the European Economic and Social Committee and the Committee of the Regions: The European Green Deal from 11 December 2019 ("European Green Deal").

[72] Hála et al. (2022), pp. 85–110.

[73] Malý et al. (2023), pp. 119–130.

[74] Cowling and Dvouletý (2023), pp. 402–419.

However, both European Commissions under the presidency of Ursula von der Leyen shared a similar vision. Namely, the second Commission of Ursula von der Leyen brought the Statement for 2024–2029 ("Statement for 2024–2029)[75] and Political Guidelines for the next European Commission 2024−2029 Europe's Choice ("New Political Guidelines")[76] with the Clean Industrial Deal are a pragmatic continuation of the Political Guidelines[77] with the European Green Deal.

In this rapidly changing unsettled scenery, there are still some constant, or at least durable and predictable, hallmarks and milestones. Recent literature has kept highlighting that sustainable development is centred around inter- and intragenerational equity and is well recognized in the EU, at least by the decade strategy Europe 2020 from 2010.[78] The European Green Deal[79] as the alpha and omega of the endeavours of the first European Commission of Ursula von der Leyen and pivotal points of EU policies and EU law,[80] was closely linked to sustainable development in the context of SDGs. The literature observed correctly that the European Green Deal impacted the understanding and application of sustainable development by all European stakeholders, regardless of their industry[81] or size.[82] The European Green Deal with "no net emissions of greenhouse gases by 2050, economic growth decoupled from resource use" along with the Fit for 55 package and European Climate Law with "55% emissions reductions below 1990 levels by 2030" anchored the environmental sustainable development at the heart of the modern European integration process[83] and this might be perceived as a "change of paradigm."[84] Since the Statement for 2024–2029 brought a pragmatic

[75] Von der Leyen U (2024) Statement at the European Parliament Plenary by President Ursula von der Leyen, candidate for a second mandate 2024–2029. Our first priority will be prosperity and competitiveness from 18 July 2024. Available from: https://neighbourhood-enlargement.ec.europa.eu/news/statement-european-parliament-plenary-president-ursula-von-der-leyen-candidate-second-mandate-2024-2024-07-18_en ("Statement for 2024-2029").

[76] Von der Leyen U (2024) Political Guidelines for the next European Commission 2024-2029: Europe´s Choice, 18 July 2024. Available from: https://commission.europa.eu/document/e6cd4328-673c-4e7a-8683-f63ffb2cf648_en ("New Political Guidelines").

[77] Von der Leyen U (2019) Political Guidelines for the next European Commission 2019–2024: A Union that strives for more. My agenda for Europe. Available from: https://www.europarl.europa.eu/resources/library/media/20190716RES57231/20190716RES57231.pdf ("Political Guidelines").

[78] COM(2010) 2020 Final Communication from the Commission: EUROPE 2020 A strategy for smart, sustainable and inclusive growth from 3 March 2010 ("Europe 2020")).

[79] COM(2019) 640 final Communication from the Commission to the European Parliament, the European Council, the Council, the European Economic and Social Committee and the Committee of the Regions: The European Green Deal from 11 December 2019 ("European Green Deal").

[80] Schunz (2022).

[81] Kowalska and Bieniek (2022), pp. 607–633.

[82] Bočková (2022), pp. 89–101.

[83] Guégen (2020).

[84] Schunz (2022).

continuation highlighting the importance of prosperity and competitiveness,[85] see the newly projected Clean Industrial Deal, the social sustainable development and even more economic sustainable development are joining the already well established environmental sustainable development. In addition, it should be emphasized that the Clean Industrial Deal was basically conceived in a bottom-up approach via the Antwerp Declaration on 20th February 2024 by 73 business leaders asking to complement the European Green Deal by a European Industrial Deal which will bring clarity, predictability and confidence in the EU and its industrial policy. The sustainability is to be multi-spectral.

European style of sustainability is conceived as a duty and potentially the focal point of policies and law and so pre-disposed to be heavily observed and commented on by academic literature addressing a myriad of branches of social sciences. However, the myriad of definitions of sustainability has been expanding rather than shrinking by consolidation. The statement "sustainability: I know it when I see it"[86] became iconic and has appeared even more fitting than ever before and its overlap with CSR and CSV kept adding ambiguity and subjectivism.[87] After some hesitations, recent literature settled for the distinction ending the decades long confusion of sustainability and sustainable development. Accordingly, sustainability was found to tackle systems while sustainability development was found to tackle human needs and well-being.[88] Similarly, after a lengthy confrontation between proponents of the idea that CSV is a revolutionary concept,[89] and their opponents, perceiving CSV as a derivation from prior theories and models such as a pyramid or CSR,[90] one decade ago, there prevailed the admission that CSV builds organically upon pre-existing thoughts.[91] This should be considered, while classifying whether sustainability and sustainable development represents a normal science development or a revolutionary science, aka a paradigm change,[92] in Europe and, perhaps, in the entire world.

[85] Von der Leyen U (2024) Statement at the European Parliament Plenary by President Ursula von der Leyen, candidate for a second mandate 2024–2029. Our first priority will be prosperity and competitiveness from 18 July 2024. Available from: https://neighbourhood-enlargement.ec.europa.eu/news/statement-european-parliament-plenary-president-ursula-von-der-leyen-candidate-second-mandate-2024-2024-07-18_en ("Statement for 2024-2029").

[86] White (2013), pp. 213–217.

[87] MacGregor Pelikánová and Sani (2023), pp. 813–851.

[88] Feil and Schreiber (2017), pp. 667–681.

[89] Wójcik (2016), pp. 32–55.

[90] Strand et al. (2015), pp. 1–15.

[91] Porter and Kramer (2014), pp. 149–151.

[92] Kuhn (1970), p. 92.

1.5 Quo Vadis?

Sustainability entered the twenty-first century while requiring addressing environmental and social aspects[93] and wanting to move from the long-term to the perpetual vision centred around goals, ideally SDGs, desired by (almost) all.[94] Sustainable development should be the process *par excellence* to reach it via a multi-stakeholder model.[95] Political science and organizational analysis bring forth relevant suggestions about adjustments of individuals and institutions, bureaucratic roles and standard operating procedures.[96] It is illustrative to point out the aphorism "where you stand depends upon where you sit."[97] There is no doubt about where the EU sits, in particular its Commission under the presidency of Ursula von der Leyen with its six priorities headed by the European Green Deal[98] and its Statement for 2024–2029 having for the priority prosperity and competitiveness headed by the Clean Industrial Deal.[99] There should no longer remain any doubt that European style of sustainability keeps resting on all three pillars. Namely, economic sustainable development needs to reconcile not only the interests of environmental and social sustainable development but, as well, needs to balance equity and fiscal soundness,[100] so that the burning question is related to conflicts about fairness pragmatism and pragmatic fairness. Social sustainability development is not about meeting the needs of everybody, but about removing barriers blocking individuals from meeting their needs,[101] so that the burning question is about the selection of these barriers and the extent of their removal in the context of many issues such as human rights, gender equity and equality, public participation and the rule of law, promotion of peace and social stability.[102] Environmental sustainability development is perhaps the most controversial and complex in the current literature. Nature and the ecosystem should maintain their integrity and the natural capital and natural resources must be harvested no faster than they can be regenerated, while waste must be emitted no

---

[93] Petera et al. (2021), pp. 1–20.

[94] MacGregor Pelikánová et al. (2024b), pp. 1–29.

[95] Panta (2019), pp. 181–190.

[96] O'Riordan and Jordan (1999), pp. 81–93.

[97] Allison (1971).

[98] Von der Leyen U (2019) Political Guidelines for the next European Commission 2019–2024: A Union that strives for more. My agenda for Europe. Available from: https://www.europarl.europa.eu/resources/library/media/20190716RES57231/20190716RES57231.pdf ("Political Guidelines").

[99] Von der Leyen U (2024) Statement at the European Parliament Plenary by President Ursula von der Leyen, candidate for a second mandate 2024–2029. Our first priority will be prosperity and competitiveness from 18 July 2024. Available from: https://neighbourhood-enlargement.ec.europa.eu/news/statement-european-parliament-plenary-president-ursula-von-der-leyen-candidate-second-mandate-2024-2024-07-18_en ("Statement for 2024-2029").

[100] Zhai and Chang (2019), pp. 369–397.

[101] Kolk (2016), pp. 23–34.

[102] Mensah (2019).

faster than it can be assimilated by the environment.[103] Hence, the burning questions are the very feasibility of the balanced priorities, as already pointed out by Meadows,[104] and who and how (in what way and intensity) is accountable or responsible or liable for them? There are no shortcuts to address these two questions, instead an extensive, but at the same time highly interesting, intellectual journey needs to be completed and this monograph is both a compass and map to do so. We cannot control the direction of the wind, but we can adjust our sails.

In that regard, this monograph is one of the first steps in doing so because the future and even the present of European style of sustainability and its regime cannot be understood without achieving a deeper appreciation of its history and conceptual roots. Therefore, after the opening of the monograph via the Introduction (1.) and identifying key sources and methods for their exploration (2.), the terminology and scientific background of sustainability (3.) along with its origins are presented (4.). Then, a multi-disciplinarity, chronological and contextual presentation of the evolution of sustainability from Antiquity to the twentieth century (5.) and the modern history of sustainability (6.) follow. Sustainability has many faces and it needs to be appreciated in a dynamic manner, see the sustainable development (7.), via a multi-stakeholder model involving private subjects via CSR (8.), and allowing the enforcement sui generis by investors via ESG (9.). This plethora of information allows to move to pioneering conclusions (10.). With a touch of exaggeration, it can be suggested that sustainability turned from the original humble and simplistic spontaneous idea into a respectable legal concept. At the beginning, it was rather a practical cogitation of leaders and autocrats regarding the local management of agricultural production with a vague content and obscure legal nature. However, over time, its meaning became broader, clearer and ultimately even general and universal.

In Europe, sustainability became a genuine concept, which is an inherent and integral part of policies and law. Indeed, policies and law have been shaping the concept of sustainability and the concept of sustainability has been shaping them, in particular their aiming and interpretation. If legal concepts are the building blocks of legal doctrine, then the concept of sustainability is a block which is growing and hardening and which can be placed anywhere. In the twenty-first century, the question is not whether sustainability is a legal concept, but rather what kind of legal concept it should be, in particular whether to continue with the multistakeholder model with mandatory reporting and standardization or to cross the Rubicon and take a more radical direction regarding regulation. The current discussion regarding issuing new policies and law instruments often jumps too quickly to conclusions about how to go for sustainability, without considering its organically developed nature and its genuine place and role in current European society. Sustainability and its legal regime deserve to be better understood.

Arguably, the proper English version of the iconic aphorism of the Spanish philosopher George Santayana from The Life of Reason: The Phases of Human

[103] Evers (2018).

[104] Meadows et al. (1972).

Progress - Reason in Common Sense (1905), vol. 1 ch. 12, p. 284 is "Those who cannot remember the past are condemned to repeat it." However, considering the current stage of the society and use of resources, we can ask if there is even a chance for repeating and having a second chance. Hence, we do not have much of a choice, for better of worse we need to know the roots and evolution of our concept of sustainability and its regime and duties—our European style of sustainability. Only after that can we engage with its current setting, interpretation and application.

Legal Documents

UN General Assembly Resolution A/RES/217(III) The Universal Declaration of Human Rights from 10 December 1948 ("UDHR")

UN Annex to document A/42/427 Report of the World Commission on Environment and Development: Our Common Future from 4 August 1987 ("Brundtland Report")

UN Conference on Environment and Development in Rio de Janeiro Declaration Agenda 21 from 3 to 14 June 1992 ("Agenda 21")

Lisbon European Council on 23 and 24 March 2000 Presidency Conclusions ("Lisbon Strategy")

COM(2010) 2020 Final Communication from the Commission: EUROPE 2020 A strategy for smart, sustainable and inclusive growth from 3 March 2010 ("Europe 2020")

Directive 2013/34/EU of the European Parliament and of the Council of 26 June 2013 on the annual financial statements, consolidated financial statements and related reports of certain types of undertakings, amending Directive 2006/43/EC of the European Parliament and of the Council and repealing Council Directives 78/660/EEC and 83/349/EEC ("Accounting Directive")

Directive 2014/95/EU of the European Parliament and of the Council of 22 October 2014 amending Directive 2013/34/EU as regards disclosure of non-financial and diversity information by certain large undertakings and groups ("NFRD")

UN General Assembly Resolution A/RES/71/1 Transforming our world: the 2030 Agenda for Sustainable Development from 25 September 2015 – People, Planet, Prosperity, Peace, and Partnership ("Agenda 2030")

Treaty on EU (C 202/1, OJ C 202 7.6.2016, p. 13) – consolidated in 2020 ("TEU")

Treaty on the Functioning of the EU (C 202/01, OJ C 202 7.6.2016, p. 47) – consolidated in 2020 ("TFEU")

Charter of Fundamental Rights of the European Union (2016/C 202/02, OJ C 202, 7.6.2016, pp. 389–405) ("Charter")

Von der Leyen U (2019) Political Guidelines for the next European Commission 2019-2024: A Union that strives for more. My agenda for Europe. Available from: https://www.europarl.europa.eu/resources/library/media/20190716RES57231/20190716RES57231.pdf ("Political Guidelines")

Regulation (EU) 2019/2088 of the European Parliament And of the Council of 27 November 2019 on sustainability-related disclosures in the financial services sector ("SFDR")

COM/2019/640 final Communication from the Commission to the European Parliament, the European Council, the Council, the European Economic and Social Committee and the Committee of the Regions: The European Green Deal from 11 December 2019 ("European Green Deal")

Regulation (EU) 2020/852 of the European Parliament and of the Council of 18 June 2020 on the establishment of a framework to facilitate sustainable investment, and amending Regulation (EU) 2019/2088 ("Taxonomy Regulation")

Regulation (EU) 2021/1119 of the European Parliament and of the Council of 30 June 2021 establishing the framework for achieving climate neutrality and amending Regulations (EC) No 401/2009 and (EU) 2018/1999 ('European Climate Law')

Directive (EU) 2022/2464 of the European Parliament and of the Council of 14 December 2022 amending Regulation (EU) No 537/2014, Directive 2004/109/EC, Directive 2006/43/EC and Directive 2013/34/EU, as regards corporate sustainability reporting ("CSRD")

Commission Delegated Regulation (EU) 2023/2772 of 31 July 2023 supplementing Directive 2013/34/EU of the European Parliament and of the Council as regards sustainability reporting standards ("ESRS Regulation")

Directive (EU) 2024/1760 of the European Parliament and of the Council of 13 June 2024 on corporate sustainability due diligence and amending Directive (EU) 2019/1937 and Regulation (EU) 2023/2859 ("CSDDD" or "CS3D")

Von der Leyen U (2024) Statement at the European Parliament Plenary by President Ursula von der Leyen, candidate for a second mandate 2024–2029. Our first priority will be prosperity and competitiveness from 18 July 2024. Available from: https://neighbourhood-enlargement.ec.europa.eu/news/statement-european-parliament-plenary-president-ursula-von-der-leyen-candidate-second-mandate-2024-2024-07-18_en ("Statement for 2024-2029")

Von der Leyen U (2024) Political Guidelines for the next European Commission 2024–2029: Europe's Choice, 18 July 2024. Available from: https://commission.europa.eu/document/e6cd4328-673c-4e7a-8683-f63ffb2cf648_en ("New Political Guidelines")

References

Allison GT (1971) Essence of decision. Little Brown and Co., Boston

Anderson TK, Kosnik LR (2002) Sustainable skepticism and sustainable development. Case Western Reserve Law Rev 53(2):439–475

Balcerzak A, MacGregor Pelikánová R (2020) Projection of SDGs in codes of ethics – case study about lost in translation? Adm Sci 10(4):95. https://doi.org/10.3390/admsci10040095

Bansal S, DesJardine MR (2014) Business sustainability: it is about time. Strateg Organ 12(1): 70–78. https://doi.org/10.1177/1476127013520265

Barbiero G (2021) Affective ecology as development of biophilia hypothesis. In: L'Astorina A, Bergami C, De Lazzari A, Falchetti E (eds) Special Issue "Scientists moving between narratives towards an ecological vision". Visions for Sustainability 16, 5575, 43–78. https://doi.org/10.13135/2384

Bočková N (2022) The role of the green deal in start-ups' approach to innovation. In: Staníčková M, Melecký L (eds) Proceedings of the 6th International Conference on European Integration 2022. VSB – Technical University of Ostrava, Ostrava, pp 89–101. https://doi.org/10.31490/9788024846057

Bowen HR (1953) Social responsibility of the Businessman. Harper & Row, New York

Brouwer R (2018) On the meaning of 'system' in the common and civil law traditions: two approaches to legal unity. Utrecht J Int Eur Law 34(1):1–11. https://doi.org/10.5334/ujiel

Carolan B (2008) The Birth of the European Union: US and UK roles in the creation of a Unified European Community. Tulsa J Comp Int Law 16(1):51–65

Carroll AB (1999) Corporate social responsibility: evolution of a definitional construct. Bus Soc 38(3):268–295. https://doi.org/10.1177/000765039903800303

Cowling M, Dvouletý O (2023) Who is brave enough to start a new business during the Covid-19 pandemic? Baltic J Manage 18(3):402–419. https://doi.org/10.1108/BJM-11-2022-0414

Cristian D, Gogan LM, Artene AE, Duran V (2015) The components of sustainable development - a possible approach. Proc Econ Finance 26:806–811. https://doi.org/10.1016/S2212-5671(15)00849-7

De Schutter O (2008) Corporate social responsibility European style. Eur Law J 14(2):203–236. https://doi.org/10.1111/j.1468-0386.2007.00411.x

Destri MLD, Minà A, Picone PM (2024) Corporate social irresponsibility and stakeholders' support: evidence from a case study. J Manage Gov 28(2):1–26. https://doi.org/10.1007/s10997-022-09639-4

Dodman M, Aillon JL, Arrobbio O, Camino E, Colucci-Gray L, Ferrara E, Folco S (2020) To connect or not to connect. Is that the question? Visions Sustain 13:3–10. https://doi.org/10.13135/2384-8677/4602

Domański B (2004) West and East in 'New Europe': the pitfalls of paternalism and a claimant attitude. Eur Urban Regional Stud 11(4):377–381. https://doi.org/10.1177/0969776404046272

Evers BA (2018) Why adopt the sustainable development goals? The case of multinationals in the Colombian coffee and extractive sector: Master Thesis Erasmus. University Rotterdam, Rotterdam

Feil A, Schreiber D (2017) Sustainability and sustainable development: unraveling overlays and scope of their meanings. Cadernos 14(3):667–681. https://doi.org/10.1590/1679-395157473

Freeman RE (2017) The new story of business: towards a more responsible capitalism. Bus Soc Rev 122(3):449–465

Friede G, Busch T, Bassen A (2015) ESG and financial performance. Aggregated evidence from more than 2000 empirical studies. J Sustain Financ Invest 5:210–233. https://doi.org/10.1080/20430795.2015.1118917

Friedman M (2007) The social responsibility of business is to increase its profits. In: Zimmerli WC, Richter K, Holzinger M (eds) Corporate ethics and corporate governance. Springer, Berlin, Heidelberg, pp 173–178

Fogarty L, Creanza N, Feldman MW (2015) Cultural evolutionary perspectives on creativity and human innovation. Trends Ecol Evol 30(12):736–754. https://doi.org/10.1016/j.tree.2015.10.004

Gössling T, Vocht C (2007) Social role conceptions and CSR policy success. J Bus Ethics 74:363–372. https://doi.org/10.1007/s10551-007-9512-3

Guégen, D (2020) Green deal: the long battles lying ahead. EurActiv. https://www.euractiv.com/section/energy-environment/opinion/green-deal-the-long-battles-lying-ahead/. Accessed 8 Nov 2023

Hála M, Cvik ED, MacGregor Pelikánová RM (2022) Logistic Regression of Czech Luxury fashion purchasing habits during the Covid-19 pandemic – old for loyalty and young for sustainability? Folia Oeconomica Stetinensia 22(1):85–110. https://doi.org/10.2478/foli-2022-0005
Harlow J, Golub A, Allenby B (2013) A review of utopian themes in sustainable development discourse. Sustain Dev 21(4):270–280. https://doi.org/10.1002/sd.522
Jones A, Sufrin B, Dunne N (2023) EU competition law. Text, cases and materials, 8th edn. Oxford University Press, Oxford
Kolk A (2016) The social responsibility of international business: from ethics and the environment to CSR and sustainable development. J World Bus 51(1):23–34. https://doi.org/10.1016/j.jwb.2015.08.010
Kowalska A, Bieniek M (2022) Meeting the European green deal objective of expanding organic farming. Equilibrium 17(3):607–633. https://doi.org/10.24136/eq.2022.021
Kramer MR, Pfitzer MV (2016) The ecosystem of shared value. Harv Bus Rev 94:80–89
Kuhn TS (1970) The structure of scientific revolutions. University of Chicago Press, Chicago
Malý M, Cvik ED, MacGregor Pelikánová R (2023) EU sanctions against the Russian Federation and their implications for the foreign trade of the Czech Republic. AGRIS on-line Pap Econ Inform 15(3):119–130. https://doi.org/10.7160/aol.2023.150310
MacGregor Pelikánová R (2017) Constantine's Christianity for the (Dis)integrated EU - Déjà vu of Constantine's Roman governance reflecting of the mistrial of Jesus for EU? Dialogo 4(1):81–98. https://doi.org/10.18638/dialogo.2017.4.1
MacGregor Pelikánová R (2024) EU legislative proposals to involve customers in the greening of the economy. Lawyer Quart 2:218–235. https://tlq.ilaw.cas.cz/index.php/tlq/article/view/597
MacGregor Pelikánová R, Nesbitt T, Balcerzak AP, Oulehla J (2024a) (In)effective communication about social responsibility? Examining large European businesses in the Czech Republic. Bus Manage Econ Eng 22(2):214–239. https://doi.org/10.3846/bmee.2024.21414
MacGregor Pelikánová R, Sani M (2023) Luxury, slow and fast fashion – a case study on the (un)sustainable creating of shared values. Equilibrium. Quart J Econ Econ Policy 18(3): 813–851. https://doi.org/10.24136/eq.2023.026
MacGregor Pelikánová R, Sani M, Rubáček F (2024b) Sustainable and responsible creation of shared values in the fast fashion industry – I Will Believe It When I See It. Visions Sustain 21(8873):1–29. https://doi.org/10.13135/2384-8677/8873
Meadows DH, Meadows DL, Randers J, Behrens WW (1972) The limits of growth. Universe Books, New York
Mensah J (2019) Sustainable development: Meaning, history, principles, pillars, and implications for human action: literature review. Cog Soc Sci 5(1). https://doi.org/10.1080/23311886.2019.1653531
O'Riordan T, Jordan A (1999) Institutions, climate change and cultural theory: towards a common analytical Framework. Glob Environ Change 9:81-93
Panta ND (2019) Clashing perspectives on sustainable development. Stud Bus Econ 14(1): 181–190. https://doi.org/10.2478/sbe-2019-0014
Petera P, Wagner J, Pakšiková R (2021) The influence of environmental strategy, environmental reporting and environmental management control system on environmental and economic performance. Energies. J Relat Sci Res Technol Dev Eng Stud Policy Manage 14(15):1–20. https://doi.org/10.3390/en14154637
Porter ME, Kramer MR (2011) CSR creating shared value – how to reinvent capitalism – and unleash a wave of innovation and growth. Harv Bus Rev 89:62–77
Porter ME, Kramer MR (2014) A response to Andrew Crane et al.'s article. Calif Manage Rev 56: 149–151
Purvis B, Mao Y, Robinson D (2019) Three pillars of sustainability: in search of conceptual origins. Sustain Sci 14(3):681–695. https://doi.org/10.1007/s11625-018-0627-5
Schunz S (2022) The 'European Green Deal' – a paradigm shift? Transformations in the European Union's sustainability meta-discourse. Polit Res Exchange 4(1). https://doi.org/10.1080/2474736X.2022.2085121

Stamper RK (1991) The role of semantics in legal expert systems and legal reasoning. Ratio Juris 4(2):219–244

Stark R (2009) God's battalions. The case for the crusades. Harper One, New York

Stark R (2014) How the West Won. ISI Books, Delaware

Strand R, Freeman E, Hockerts K (2015) Corporate social responsibility and sustainability in Scandinavia: an overview. J Bus Ethics 127(1):1–15. https://doi.org/10.1007/s10551-014-2224-6

Tóthová M (2023) Several aspects of the comparative method in law. Lawyer Quart 13(1):78–88

Van Tulder R, Keen N (2018) Capturing collaborative challenges: designing complexity-sensitive theories of change for cross-sector partnerships. J Bus Ethics 150:315–332. https://doi.org/10.1007/s10551-018-3857-7

Van Tulder R, Rodrigues SB, Mirza H et al (2021) The UN's sustainable development goals: can multinational enterprises lead the decade of action? J Inter Bus Policy 4:1–21. https://doi.org/10.1057/s42214-020-00095-1

Van Tulder R, Van Mil E (2023) Principles of sustainable business frameworks for corporate action on the SDGs. Routledge, London

White MA (2013) Sustainability: I know it when I see it. Ecol Econ 86:213–217. https://doi.org/10.1016/j.ecolecon.2012.12.020

Wójcik P (2016) How creating shared value differs from corporate social responsibility. J Manage Bus Adm Central Europe 24(2):32–55. https://doi.org/10.7206/jmba.ce.2450-7814.168

Zhai T, Chang Y (2019) Standing of environmental public-interest litigants in China: evolution, obstacles and solutions. J Environ Law 30:369–397. https://doi.org/10.1093/jel/eqy011

Zatonatska T, Soboliev O, Zatonatskiy D, Dluhopolska T, Rutkowski M, Rak N (2024) A comprehensive analysis of the best practices in applying environmental, social, and governance criteria within the energy sector. Energies 17:2950. https://doi.org/10.3390/en17122950

Chapter 2
Materials and Methods

The term *sustainability* is an abstract, ephemeral and evolving omnipresent expression with a growing significance and reaching even the strength and intensity of an enforceable command. Since it is the subject matter of various branches of social sciences, it leads to multidisciplinary research and studies built upon heterogenous sources and data and explored via a myriad of predominantly qualitative methods and mechanisms. These endeavours are centred around two key questions: (Sect. 2.1) searching for the foundation, development and meaning of sustainability (Sect. 2.2) and searching for its regime (Sect. 2.3). This inevitably leads to the revelation of a challenging tension between expectations and reality. A multitude of materials should be explored while using proper methods in order to overcome this tension and to tie together the ends (Sect. 2.4).

2.1 Sustainability European Style: Two Questions to Be Addressed

The roots and concepts as well as the evolution and the current meaning of sustainability are perceived dramatically on the global level,[1] on the level of the EU[2] and even the level of EU member states.[3] As already mentioned, sustainability is about the use of resources, in particular tangible resources, and the judgment and balancing in their use along with concerns regarding their exhausting in a selfish

[1] White (2013), pp. 213–217.

[2] De Schutter (2008), pp. 203–236.

[3] Domański (2004), pp. 377–381.

R. MacGregor Pelikánová, *Sustainability in Europe*,
https://doi.org/10.1007/978-3-031-74689-5_2

manner[4] and the eternal temptation[5] to be selfish[6] and for a reckless immediate self-gratification.[7] This inevitably leads to a set of cultural differences and so basically the same issue (how to use tangible, and so inevitably exhaustible, resources) is addressed differently in different societies in different eras. As a matter of fact, it is addressed differently even by different groups in the same society in the same era. The perception and attitude of sustainability is linked to responsibility, planning, trust and many other considerations. Western civilization based on and shaped by Christianity has developed for many centuries predominantly in Europe and resulted in the current concept of sustainability in Europe. Naturally, the issue of exhausting tangible resources has acquired a global dimension and it would be remiss to ignore the existence and development of the concept dealing with it outside Europe, such as in Asia or Africa, and based on foundations other than Christianity, such as Islam, Confucianism, and Buddhism, and strongly influenced by other factors such as colonialism.[8]

Well, the global discourse about the concept of sustainability is far from reaching a consensus and even in Europe there are discrepancies. Nevertheless, there is a mutual understanding in Europe that sustainability should not be reduced to an ambiguous and/or controversial concept,[9] but instead it has to be a real meta-consideration and meta-compass firmly resting on three pillars.[10] The EU even progressively moves towards establishing not only pro-sustainable policies and strategies, see Europe 2020,[11] Political Guidelines[12] with the European Green Deal[13] and New Political Guidelines[14] with the Clean Industrial Deal, but as well to enacting mandatory provisions about the legal duty to report about sustainability and even the duty to engage in sustainability. This can be demonstrated through the

[4]Barbiero (2021), pp. 43–78.

[5]MacGregor Pelikánová et al. (2024), pp. 1–29.

[6]D'Adamo and Gastaldi (2023), p. 140944.

[7]Ding et al. (2021), p. e0256966.

[8]Kwashirai (2013), pp. 166–196.

[9]MacGregor Pelikánová et al. (2024), pp. 1–29.

[10]Purvis et al. (2019), pp. 681–695.

[11]COM(2010) 2020 Final Communication from the Commission – EUROPE 2020 A strategy for smart, sustainable and inclusive growth from 3 March 2010 ("Europe 2020").

[12]Von der Leyen U (2019) Political Guidelines for the next European Commission 2019–2024: A Union that strives for more. My agenda for Europe. Available from: https://www.europarl.europa.eu/resources/library/media/20190716RES57231/20190716RES57231.pdf("Political Guidelines").

[13]COM/2019/640 final Communication from the Commission to the European Parliament, the European Council, the Council, the European Economic and Social Committee and the Committee of the Regions: The European Green Deal from 11 December 2019 ("European Green Deal").

[14]Von der Leyen U (2024) Political Guidelines for the next European Commission 2024–2029: Europe's Choice, 18 July 2024. Available from: https://commission.europa.eu/document/e6cd4328-673c-4e7a-8683-f63ffb2cf648_en ("New Political Guidelines").

legislative history of the accounting directives from the NFRD[15] to the CSRD[16] and leading the current ESG framework entailing CSR in selected sectors, such as in banking via the SFDR,[17] as well as generally, see the Taxonomy Regulation.[18]

It can even be argued that sustainability has reached, at least in some jurisdictions, the stage of a well-recognized legal concept generating obligation and thus giving rights and imposing duties upon a large heterogenous pool of subjects, perhaps even all of us. The EU and EU member states are such jurisdictions and so sustainability in Europe is perhaps an ephemeral but definitely not illusory concept, it is the reality and even this reality is explicitly covered by ethics and law, which create duties and impose responsibilities[19] and even liabilities. At the same time, if sustainability is to be real and not illusory in Europe, then its meaning and regime must be understood and endorsed by the entire society. Manifestly, the current situation in Europe does not meet such expectations. Shortcomings regarding the understanding of sustainability, and its regime, erode and undermine the undeniable drive and commitment.[20] There are very few commands enjoying a consensual support and these few include the rejection of waste and the desire for justice and fairness.[21] Sustainability is against waste, while a misunderstood sustainability regime is waste. Sustainability is based on respect, while a misunderstood sustainability regime erodes the values-based society. Often, trust is mentioned as the key word,[22] which should be the glue and causality link.[23]

Undoubtedly, sustainability is very much needed. However, since there is not a consensus about the definition[24] nor the interpretation nor the application of sustainability[25] and about its implied duties and their nature (moral, legal, political), thus the entire drive for sustainability is often misinterpreted and keeps losing momentum. Despite many academic, professional and laic publication endeavours, it remains a vacuum to be filled. Therefore, the research aim of this monograph is to

---

[15]Directive 2014/95/EU of the European Parliament and of the Council of 22 October 2014 amending Directive 2013/34/EU as regards disclosure of non-financial and diversity information by certain large undertakings and groups ("NFRD").

[16]Directive (EU) 2022/2464 of the European Parliament and of the Council of 14 December 2022 amending Regulation (EU) No 537/2014, Directive 2004/109/EC, Directive 2006/43/EC and Directive 2013/34/EU, as regards corporate sustainability reporting ("CSRD").

[17]Regulation (EU) 2019/2088 of the European Parliament and of the Council of 27 November 2019 on sustainability-related disclosures in the financial services sector ("SFDR").

[18]Regulation (EU) 2020/852 of the European Parliament and of the Council of 18 June 2020 on the establishment of a framework to facilitate sustainable investment, and amending Regulation (EU) 2019/2088 ("Taxonomy Regulation").

[19]Kolk (2016), pp. 23–34.

[20]Bansal and DesJardine (2014), pp. 70–78.

[21]MacGregor Pelikánová et al. (2024), pp. 1–29.

[22]Eger et al. (2019), pp. 141–153.

[23]Heckman (2005), pp. 1–97.

[24]White (2013), pp. 213–217.

[25]MacGregor Pelikánová et al. (2024), pp. 1–29.

explore its roots and evolution in order to contextually figure out the development of the meaning and the regime of sustainability in Europe, aka Sustainability European Style. Such a deeper and contextual appreciation of the foundations and developing setting should assist with the current needs related to the interpretation and application of the duties generated by the sustainability regime based on EU policies and law. To put it differently, this monograph has two goals, each based upon one question and leading to an analysis bringing clarification and enhancing awareness and, hopefully, inspiring commitments. These two questions and the following analysis are about: (i) what is the foundation, development and meaning of the concept of sustainability in Europe and (ii) what kind of regime and duties has it involved in Europe? The materials and methods to be employed in this explorative undertaking are inherently determined by these two questions about the definition and the regime.

2.2 The Search for the Foundation, Development and Meaning of the Concept of Sustainability

Regarding the definition and meaning of the concept of sustainability in the current EU, it is necessary to go back in time. Namely, it is necessary to explore the roots and evolution of this concept in order to contextually figure out the denotation of sustainability in the current EU. It must be recognized that sustainability is an English word without Anglo-Saxon roots, which has been adopted by the multilingual EU policy and law. Consequently, the (modern) English term sustainability, as such, needs to be linguistically analysed while paying attention to the dynamics and particularities of legal and professional English and their varied origins combining Anglo-Saxon, Latin, Norman, French, Scandinavian and other terminologies.[26] Further, the (modern) EU term sustainability, as such, needs to be linguistically analysed while considering that EU language, which is the result of an artificial a priori choice. Thus, sustainability is becoming a neologism to overcome national diversities.[27]

Therefore, the materials and methods to be employed to study the past, present and future of sustainability in Europe must reflect the obvious, but often underplayed and underestimated, fact that, for centuries, and even millennia, sustainability has been mentioned and referred to (often heavily) by different authorities and different sources from different eras[28] in varying jurisdictions and in different languages. Regarding the roots, evolution and understanding of sustainability in Europe, in particular its current meaning, it must be emphasized that the term sustainability has undergone a millennia long evolution and is at the centre of the intersection of three

[26] Schneiderová (2018), pp. 117–126.

[27] Ioriatti (2022), pp. 275–291.

[28] White (2013), pp. 213–217.

dimensions—environmental, social and economic. All of them are covered by the social sciences which deal with human behaviour in its social and cultural aspects. Therefore, sustainability is a key term from, and for, a number of social science branches such as economics (the study of the production, distribution, and consumption of goods and services in a sustainable manner), law (the study of legal systems as frameworks to interpret and use the law), political science, etc. Thus, social science materials and methods are, *prima facie,* suitable for the understanding, interpretation and application of the term sustainable. A number of branches of social science, such as philosophy or linguistics, even assist its sister branches of social science, such as economics or law, in identifying proper sources, materials and scientific methods. Plainly, since sustainability is about human behaviour addressed by different branches of social science, thus its study must reflect the heterogeneity of these branches, as well as their common point (human behaviour), and apply it in order to discover its meaning.[29]

Materials to explore the roots and evolution of sustainability in Europe must be heterogenous, multi-jurisdictional and predominantly from academic sources[30] followed by policy and legislative sources dating from the Biblical era until now. There is no primary source of materials and the inherent randomness of the material selection is to be reduced by the broad reach of heavily referred sources, such as publications from the Web of Science aka WoS and Scopus databases which are heavily quoted and which cover sustainability in the European context. Particular attention is to be paid to materials tied to crises, because typically crises accelerate pre-existing trends and are both challenges and opportunities for change,[31] see e.g., the impact of Covid-19.[32] It has been arguably advanced even by Albert Einstein that a crisis is a necessary stimulus for human progress, inventiveness and innovations.[33]

Methods to explore the roots and evolution of sustainability in Europe are obviously holistic, qualitative and comparative. Comparative studies in the law and legal system arena entail the functionalist principle (how a similar problem is addressed in various jurisdictions), analytical method, context-oriented method, causality focussed method[34] and, perhaps most importantly, the historical method.[35]

The exploration of the history (predominantly based on literature) of the term leading to its current meaning needs to address the contextual setting and the dynamics of interpretations matching the given sources. Plainly, the processing line will start with a qualitative critical content analysis[36] of literature sources from different jurisdictions and from different branches of social science about the

---

[29] Feil and Schreiber (2017), pp. 667–681.

[30] Mensah (2019).

[31] Kovoor-Misra (2009), pp. 494–510.

[32] Hála et al. (2022), pp. 85–110.

[33] D'Adamo and Lupi (2021), p. 1861.

[34] Heckman (2005), pp. 1–97.

[35] Tóthová (2023), pp. 78–88.

[36] Krippendorff (2003).

roots of sustainability[37] and then proceed to the comparative study of the evolution of its meaning. Both the qualitative contextual content analysis[38] and critical comparison will be done predominantly manually, but still, complementary computerized language analysis techniques,[39] such as LIWC, are to be included and (with caution) used. Since qualitative research rarely follows a smooth trajectory from hypothesis to findings, the critical endeavour will be about the open-minded "zig-zagging" research and interpretation of these sources[40] and the selection and employment of appropriate linguistic and functionalist tools.[41] The resulting propositions are to be refreshed by glossing[42] and Socratic questioning.[43]

Manifestly, the binary true-false propositional logic is not appropriate for the exploration of such sources and materials. However, even the variables-based predicate logic could hardly be employed automatically and, *per se*, due to the presence, if not predominance, of argumentative features of involved social sciences. The starting point in this definition endeavour is to go back and appreciate the origins, roots and evolution of sustainability in a multidisciplinary and multijurisdictional manner. Semantics, which is a branch of linguistics, studies the meaning in language in general, as well as in special settings, such as a legal setting.[44] The legal science continuously systematized its study, analysis and resulting knowledge about the law,[45] while they both use the same ambiguous, vague and abstract instruments—words. Consequently, the law and legal science are much more argumentative than axiomatic and they need the employment of legal-setting oriented semantics with a number of universal approaches and rules (literal, etc.), of special branches approaches and rules (golden, mischief and purposive), and even regionally particular approaches and rules (teleological due to the CJ EU determination to go for the spirit of the EU law via an autonomous pro-integration interpretation). Naturally, the semantic analysis is not detached from other linguistic analyses and it builds upon the word structure analysis (morphology) and the sentence structure analysis (syntax), while the evolving context and social situation appreciation (pragmatism) is critical. Considering the term sustainability and its position in social sciences, the use of logic as a mechanical model of language should give the priority to a more open, holistic, flexible, pragmatic and causality-oriented search for a meaningful understanding, interpretation and application. Finally, the term sustainability is clearly within the reach of another social science branch—philosophy, in particular its discipline focusing on the nature and

---

[37] Kuckartz (2014).

[38] Silverman (2013).

[39] Boyd (2017), pp. 161–189.

[40] Silverman (2013).

[41] Tóthová (2023), pp. 78–88.

[42] Hyland (2007), pp. 266–285. https://doi.org/10.1093/applin/amm011.

[43] Areeda (1996), pp. 911–922.

[44] Stamper (1991), pp. 219–244.

[45] Tóthová (2023), pp. 78–88.

organization of reality, i.e. ontology. Therefore, the research design regarding the meaning of sustainability is more exploratory than conclusive.[46]

2.3 The Search for the Regime and Duties Related to the Evolving Concept of Sustainability

Regarding the regime of sustainability in Europe, in particular the dynamics of a set of duties reflecting the accountability and/or responsibility and/or liability, it is necessary to focus on current materials and methods complementarily boosted by a historical and linguistic exploration. This is clearly intra-related to the issue of the evolving meaning of sustainability, because the regime matches developmental stages and the social context, economics, culture, religion, etc.[47] The exploration and analysis will be based more on law than on literature and the processing will be mainly reflecting the methodology for legal studies with a strong focus on the law interpretation approaches, such as the teleological regarding the EU law.

Materials to explore the regime of sustainability, in particular the generated duties and their nature, are rather homogenous, predominately from the European jurisdictions and principally from legislative sources complemented by policy and academic sources. The primary source of materials are law databases, historical law collections and the EU domain EurLex and Curia, along with the official Internet website of the European Commission. Clearly, modern European integration leads to the need for a careful consideration of the law and policy dynamics while following the motto "united in diversity" and the heavily quoted *l'effet utile.* The secondary source of materials will be academic publications tied to these documents from EurLex and national legislative and academic sources from EU member states. However, the EU regime of sustainability, and duties generated by it, are not only the products of European endeavours and the common law contribution, but also the involvement of the UN needs to be considered. In particular, the current EU regime of sustainability, both on legislative and policy levels, is strongly shaped and influenced by several UN instruments and it cannot be overemphasized that the European Commissions of Jean-Claude Juncker (2014–2019) and Ursula von der Leyen (2019–2024) have fully endorsed and closely followed the Agenda 2030[48] with its 17 SDGs and 169 targets.[49] Therefore, similar to the definition and its meaning, even the current regime of sustainability in the EU is the product of a particular evolution and thus its

[46] Yin (2008).

[47] Tóthová (2023), pp. 78–88.

[48] UN General Assembly Resolution A/RES/71/1 Transforming our world: the 2030 Agenda for Sustainable Development from 25 September 2015—People, Planet, Prosperity, Peace, and Partnership ("Agenda 2030").

[49] Balcerzak and MacGregor Pelikánová (2020), p. 95.

proper understanding requires a contextual and multidisciplinary study based on an appropriate methodology.

Methods to explore the regime of sustainability, in particular the generated duties and their nature, are chiefly interpretative and auxiliary text analyses. These methods are to be applied based on the used sources, their type, time and jurisdictional origin. Highly important is the exploration of historical and international law sources as well as of two key lines of recent EU legislation: (i) the older, general and rather procedural line entailing sustainability reporting based on Directives extending from the NFRD[50] to the CSRD[51] along with the terminology setting by the Taxonomy Regulation,[52] (ii) the newer special and rather substantive line about a particular sectorial setting for the financial and banking industry,[53] for recycling, for labelling, for the environment, etc., see the SFDR.[54] Naturally, these are currently further expanded in the spheres of green claims v green washing,[55] QR-code labelling,[56] etc. However, this is no longer the past, this is the present of sustainability in Europe.

Undoubtedly, both the content analysis and interpretation of these legislative instruments need to follow the literal, golden rule, mischief, purposive and teleological approaches and recognize the judicial authorities, such as the CJ EU as well as other authorities. However, considering the novelty, and thus not yet developed case law, the importance of policy instruments as guidelines for the interpretation and application of these legislative instruments needs to be emphasized. The European Commission of Ursula von der Leyen brought for 2019–2024 her Political Guidelines[57] with six priorities, one of which is the European Green Deal aka EGD[58] a key

[50] Directive 2014/95/EU of the European Parliament and of the Council of 22 October 2014 amending Directive 2013/34/EU as regards disclosure of non-financial and diversity information by certain large undertakings and groups ("NFRD").

[51] Directive (EU) 2022/2464 of the European Parliament and of the Council of 14 December 2022 amending Regulation (EU) No 537/2014, Directive 2004/109/EC, Directive 2006/43/EC and Directive 2013/34/EU, as regards corporate sustainability reporting ("CSRD").

[52] Regulation (EU) 2020/852 of the European Parliament and of the Council of 18 June 2020 on the establishment of a framework to facilitate sustainable investment, and amending Regulation (EU) 2019/2088 ("Taxonomy Regulation").

[53] Rubáček et al. (2023), pp. 123–143.

[54] Regulation (EU) 2019/2088 of the European Parliament and of the Council of 27 November 2019 on sustainability-related disclosures in the financial services sector ("SFDR").

[55] MacGregor Pelikánová and Rubáček (2022), pp. 173–195.

[56] MacGregor Pelikánová (2024), pp. 218–223.

[57] Von der Leyen U (2019) Political Guidelines for the next European Commission 2019–2024: A Union that strives for more. My agenda for Europe. Available from: https://www.europarl.europa.eu/resources/library/media/20190716RES57231/20190716RES57231.pdf("Political Guidelines").

[58] COM/2019/640 final Communication:The European Green Deal from 11 December 2019 ("-European Green Deal") COM(2019) 640 final Communication from the Commission to the European Parliament, the European Council, the Council, the European Economic and Social Committee and the Committee of the Regions: The European Green Deal from 11 December 2019 ("European Green Deal").

strategic policy instrument,[59] which has nine sub-strategies, such as the EGD3 New circular economy action plan aka CEAP,[60] the EGD8 Textile strategy and the EGD9 Zero pollution action plan. The focus on these six priorities is not dramatically changed for 2024–2029 by the New Political Guidelines,[61] but of course a more pragmatic touch is noticeable, see the transformation of the European Green Deal into Clean Industrial Deal.

These six priorities, strategies and sub-strategies are both the transmission foundations between Agenda 2030 and the current EU regime for sustainability and guidelines for interpretation and application.[62] In addition, they are trend indicatives and their qualitative content analysis, along with field and longitudinal observations,[63] suggests that informal individual accountability is already, by the operation of EU law and policies, transformed into a moral responsibility with several legal liability aspects. Thusly, the interpretation and application of any instruments about sustainability in the current EU need to methodologically recognize the shift of the pendulum towards a more and more specific legal duty imposed on a growing pool of subjects.

2.4 Sustainability European Style: Tying Ends Together

Sustainability European Style is a concept generating progressively more duties to a larger pool of subjects. Although various, and not always inherently European, definitions are used, there are shared ideas about sustainability as a more than mere local and temporary suggestion. Namely, sustainability is manifestly an ongoing issue addressing the needs of various groups and societies and even various generations. Many policy and law instruments are emerging and even more initiatives are taken—some are effective, efficient and legitimate, while others less or not at all.[64]

During the last few decades, the EU has kept on issuing policy and law documents pursuant to the international law trends, to which the EU contributes, such as Agenda 2030. The EurLex and European Commission website, as well as numerous

[59] Schunz (2022).

[60] COM(2020) 98 final Communication from the Commission to the European Parliament, the Council, the European Economic and Social Committee and the Committee of the Regions: A New Circular Economy Action Plan - For a cleaner and more competitive Europe from 11 March 2020 ("CEAP").

[61] Von der Leyen U (2024) Political Guidelines for the next European Commission 2024–2029: Europe's Choice, 18 July 2024. Available from: https://commission.europa.eu/document/e6cd4328-673c-4e7a-8683-f63ffb2cf648_en("New Political Guidelines").

[62] Kowalska and Bieniek (2022), pp. 607–633.

[63] Gold (1958), pp. 217–223.

[64] Kathayat (2022), pp. 701–714.

academic publications in the EU, allows for the detection of key information sources about the term sustainability and the regime generated along with it.

However, identifying the term and regime with duties would be futile without their interpretation considering the history, UN documents such as Agenda 2030 and the EU policies and law. They must be interpreted via the teleological approach and based on deductive reasoning with an interpretation by an agent based on the legal positivism of Hart's will, reach the compromises of Hobbes´ social contract, the utilitarian concession of Bentham's pragmatism, the deontological principles of Kant's judgment and the universalism of Aristotles' law of nature remains. Consequently, subjective ethical features reflecting one's personal morals are inevitable and ultimately projected in the conduct of consumers and businesses,[65] which are to engage with SDGs.[66] Indeed, the ontological question about what is reality is both the alpha and omega of any discourse about the ephemeral sustainability duty. Such a discourse and ultimate search for its understanding, interpreting and applying needs to be done based on heterogenous sources in a methodologically organized designed manner. Given the focus of this monograph and the inherently implied need for an advanced economic, legal and semantic exploration with contextual as well as evolutionary features, the multidisciplinary search for data and their processing will be holistically done via a multi-step critical and comparative content analysis with a strong qualitative text focus.[67] The academic robustness will be increased by the juxtaposition of proposed indices about the deeper understanding, interpretation and application of the Sustainability European Style and the implied duty by their Socratic questioning and glossing based on multi-spectral field observations.[68] A methodologic processing is obviously chronological, from general to special, from concepts to regimes.

In sum, based on the research, a qualitative content analysis of literature sources about the roots of sustainability and, along with the comparative study of the evolution of its meaning, a deeper semantic and contextual understanding will be achieved. This will allow for moving on to explore more legislative, policy and academic sources about its general and particular regime and focus on its identification, interpretation and even application, including the consideration of casuistic case studies about its application in specific industries, such as the fashion industry.[69] Such a battery of sources and tools employed in the manner definition-interpretation-application should offer various perspectives and ultimately allow for a deeper appreciation of the philosophical, semantic, economic and legal dimension of Sustainability European Style. Although the concept has an ephemeral and dynamically changing nature and perhaps ambiguous and evolving meaning, Sustainability European Style is an undeniable attempt to overcome short-termism and

---

[65] MacGregor Pelikánová et al. (2021), pp. 973–1009.

[66] Van Tulder and Van Mil (2023), pp.106–119.

[67] Vourvachis and Woodward (2015), pp. 166–195.

[68] Hyland (2007), pp. 266–285.

[69] MacGregor Pelikánová and Sani (2023), pp. 813–851.

waste and to achieve fairness and solidarity between stakeholders from various jurisdictions and various generations by turning their accountability into responsibility and even into a liability. Ready or not, for better or worse, Sustainability European Style is an integral part of our life. It is our present which we can understand, appreciate and develop into the future only if we know its history and conceptual roots. The past of the Sustainability European Style is definitely not to be passed over … it is *conditio sine qua non* to engage with its current setting, interpretation and application.

Legal Documents

COM(2010) 2020 Final Communication from the Commission – EUROPE 2020 A strategy for smart, sustainable and inclusive growth from 3 March 2010 ("Europe 2020")

Directive 2014/95/EU of the European Parliament and of the Council of 22 October 2014 amending Directive 2013/34/EU as regards disclosure of non-financial and diversity information by certain large undertakings and groups ("NFRD")

UN General Assembly Resolution A/RES/71/1 Transforming our world: the 2030 Agenda for Sustainable Development from 25 September 2015 – People, Planet, Prosperity, Peace, and Partnership ("Agenda 2030")

Regulation (EU) 2019/2088 of the European Parliament and of the Council of 27 November 2019 on sustainability-related disclosures in the financial services sector ("SFDR")

Von der Leyen U (2019) Political Guidelines for the next European Commission 2019-2024: A Union that strives for more. My agenda for Europe. Available from: https://www.europarl.europa.eu/resources/library/media/20190716RES57231/20190716RES57231.pdf ("Political Guidelines")

COM/2019/640 final Communication from the Commission to the European Parliament, the European Council, the Council, the European Economic and Social Committee and the Committee of the Regions: The European Green Deal from 11 December 2019 ("European Green Deal")

COM(2020) 98 final Communication from the Commission to the European Parliament, the Council, the European Economic and Social Committee and the Committee of the Regions: A New Circular Economy Action Plan - For a cleaner and more competitive Europe from 11 March 2020 ("CEAP")

Regulation (EU) 2020/852 of the European Parliament and of the Council of 18 June 2020 on the establishment of a framework to facilitate sustainable investment, and amending Regulation (EU) 2019/2088 ("Taxonomy Regulation")

Directive (EU) 2022/2464 of the European Parliament and of the Council of 14 December 2022 amending Regulation (EU) No 537/2014, Directive 2004/109/EC, Directive 2006/43/EC and Directive 2013/34/EU, as regards corporate sustainability reporting ("CSRD")

Von der Leyen U (2024) Political Guidelines for the next European Commission 2024-2029: Europe´s Choice, 18 July 2024. Available from: https://commission.europa.eu/document/e6cd4328-673c-4e7a-8683-f63ffb2cf648_en ("New Political Guidelines")

References

Areeda PE (1996) The Socratic method. Harv Law Rev 109(5):911–922

Balcerzak A, MacGregor Pelikánová R (2020) Projection of SDGs in codes of ethics—case study about lost in translation? Adm Sci 10(4):95. https://doi.org/10.3390/admsci10040095

Bansal S, DesJardine MR (2014) Business sustainability: it is about time. Strateg Organ 12(1): 70–78. https://doi.org/10.1177/1476127013520265

Barbiero G (2021) Affective ecology as development of biophilia hypothesis. In: L'Astorina A, Bergami C, De Lazzari A, Falchetti E (eds) Special Issue "Scientists moving between narratives towards an ecological vision". Vis Sustain 16, 5575, 43–78. https://doi.org/10.13135/2384

Boyd RL (2017) Psychological text analysis in the digital humanities. In: Hai-Jew S (ed) Data analytics in digital humanities. Springer International Publishing, pp 161–189. https://doi.org/10.1007/978-3-319-54499-1_7

D'Adamo I, Gastaldi M (2023) Monitoring performance of sustainable development goals in the Italian regions. Sustainability 15:14094. https://doi.org/10.3390/su151914094

D'Adamo I, Lupi G (2021) Sustainability and resilience after COVID-19: a circular premium in the fashion industry. Sustain For 13(4):1861. https://doi.org/10.3390/su13041861

De Schutter O (2008) Corporate social responsibility European style. Eur Law J 14(2):203–236. https://doi.org/10.1111/j.1468-0386.2007.00411.x

Ding N, Frohnwieser A, Miller R, Clayton NS (2021) Waiting for the better reward: comparison of delay of gratification in young children across two cultures. PLoS One 16(9):e0256966. https://doi.org/10.1371/journal.pone.0256966

Domański B (2004) West and east in 'new Europe': the pitfalls of paternalism and a claimant attitude. Eur Urban Region Stud 11(4):377–381. https://doi.org/10.1177/0969776404046272

Eger C, Miller G, Scarles C (2019) Corporate philanthropy through the lens of ethical subjectivity. J Bus Ethics 156(1):141–153. https://doi.org/10.1007/s10551-017-3551-1

Feil A, Schreiber D (2017) Sustainability and sustainable development: unraveling overlays and scope of their meanings. Cadernos 14(3):667–681. https://doi.org/10.1590/1679-395157473

Gold RL (1958) Roles in sociological field observations. Soc Forces 36(3):217–223

Hála M, Cvik ED, MacGregor Pelikánová RM (2022) Logistic regression of Czech luxury fashion purchasing habits during the Covid-19 pandemic—old for loyalty and young for sustainability? Folia Oeconomica Stetinensia 22(1):110. https://doi.org/10.2478/foli-2022-0005

Heckman JJ (2005) The scientific model of causality. Sociol Methodol 35(1):1–97. https://doi.org/10.1111/j.0081-1750.2006.00164.x

Hyland K (2007) Applying a gloss: exemplifying and reformulating in academic discourse. Appl Linguis 28(2):266–285. https://doi.org/10.1093/applin/amm011

Ioriatti E (2022) Common contexts of meaning in the European legal setting: opening Pandora's box? Int J Semiot Law 36:275–291. https://doi.org/10.1007/s11196-022-09919-0

Kathayat A (2022) A case against CSR initiatives. Open J Bus Manage 10(2):701–714. https://doi.org/10.4236/ojbm.2022.102039

Kolk A (2016) The social responsibility of international business: from ethics and the environment to CSR and sustainable development. J World Bus 51(1):23–34. https://doi.org/10.1016/j.jwb.2015.08.010

Kovoor-Misra S (2009) Understanding perceived organizational identity during crisis and change: a threat/opportunity framework. J Organ Chang Manag 22(5):494–510. https://doi.org/10.1108/09534810910983460

Kowalska A, Bieniek M (2022) Meeting the European green deal objective of expanding organic farming. Equilibrium 17(3):607–633. https://doi.org/10.24136/eq.2022.021

Krippendorff K (2003) Content analysis: an introduction to its methodology. Sage Publications, Thousand Oaks

Kuckartz U (2014) Qualitative text analysis: a guide to methods, practice and using software. Sage Publications, Thousand Oaks

Kwashirai VC (2013) Environmental change, control and management in Africa. Glob Environ 12: 166–196

MacGregor Pelikánová R (2024) EU legislative proposals to involve customers in the greening of the economy. Lawyer Quart 2:218–235. https://tlq.ilaw.cas.cz/index.php/tlq/article/view/597

MacGregor Pelikánová R, Rubáček F (2022) Taxonomy for transparency in non-financial statements—clear duty with unclear sanction. Danube 13(3):173–195. https://doi.org/10.2478/danb-2022-0011

MacGregor Pelikánová R, Sani M (2023) Luxury, slow and fast fashion—a case study on the (un)sustainable creating of shared values. Equilibrium. Quart J Econ Econ Policy 18(3): 813–851. https://doi.org/10.24136/eq.2023.026

MacGregor Pelikánová R, MacGregor RK, Černek M (2021) New trends in codes of ethics: Czech business ethics preferences by the dawn of COVID-19. Oeconomia Copernicana 12(4): 973–1009. https://doi.org/10.24136/oc.2021.032

MacGregor Pelikánová R, Sani M, Rubáček F (2024) Sustainable and responsible creation of shared values in the fast fashion industry—I will believe it when I see it. Vis Sustain 21(8873):1–29. https://doi.org/10.13135/2384-8677/8873

Mensah J (2019) Sustainable development: meaning, history, principles, pillars, and implications for human action: literature review. Cogent Soc Sci 5(1). https://doi.org/10.1080/23311886.2019.1653531

Purvis B, Mao Y, Robinson D (2019) Three pillars of sustainability: in search of conceptual origins. Sustain Sci 14(3):681–695. https://doi.org/10.1007/s11625-018-0627-5

Rubáček F, MacGregor Pelikánová R, MacGregor RK (2023) The sustainability of ESAs triumvirate for sustainability-related disclosures in the financial sector—all for one and one for all. Acta Universitatis Lodziensis Folia Iuridica 105:123–143. https://doi.org/10.18778/0208-6069.105.08

Schneiderová A (2018) Historical background to English legal language. J Mod Sci 2(37):117–126. https://doi.org/10.13166/jms/92522

Schunz S (2022) The 'European Green Deal'—a paradigm shift? Transformations in the European Union's sustainability meta-discourse. Polit Res Exchange 4(1). https://doi.org/10.1080/2474736X.2022.2085121

Silverman D (2013) Doing qualitative research—a practical handbook, 4th edn. Sage, London. https://www.researchgate.net/publication/279187183_Doing_Qualitative_Research_A_Practical

Stamper RK (1991) The role of semantics in legal expert systems and legal reasoning. Ratio Juris 4(2):219–244

Tóthová M (2023) Several aspects of the comparative method in law. Lawyer Quart 13(1):78–88

Van Tulder R, Van Mil E (2023) Principles of sustainable business frameworks for corporate action on the SDGs. Routledge, New York

Vourvachis P, Woodward T (2015) Content analysis in social and environmental reporting research: trends and challenges. J Appl Acc Res 16(2):166–195. https://doi.org/10.1108/JAAR-04-2013-0027

White MA (2013) Sustainability: I know it when I see it. Ecol Econ 86:213–217. https://doi.org/10.1016/j.ecolecon.2012.12.020

Yin R (2008) Study research. Design methods. Sage Publications, Thousand Oaks

Chapter 3
Terminology and Scientific Background of Sustainability

Sustainability is a phenomenon with both a dynamic terminology and complex background, shaped predominantly in the sphere of social science (Sect. 3.1). Since it is both expressed and explained by language units dealing with reality and fairness, its elementary foundation is to be understood in the field of linguistics (Sect. 3.2) and appreciated in philosophy (Sect. 3.3). Indeed, its multi-disciplinary features suggest the meta dimension, i.e. the meta-sustainability (Sect. 3.4).

3.1 Sustainability in Social Science

The term sustainability is heavily used in the current global society, in particular in the EU. Ever since 2019, almost all EU policies and laws, directly or indirectly, refer to it in order to boost their justification and legitimacy. Naturally, this is projected in academic discourses as well as in discussions within the society at large and progressively mushrooms into the literature. Although the importance and massive use of the term sustainability cannot be overemphasized, it is obvious that it is an abstract, heterogenous and conceptually unsettled argumentative command. Plainly, its roots go back to ancient philosophy and its setting, they are partially obscure and vague and their interpretation is extremely diversified, often reduced to empirical observations. Consequently, the meaning, the interpretation and the application of sustainability is ephemeral and subject to a set of semantic pitfalls in all three spheres—environmental, social and economic. All of them are covered by the social sciences which deal with human behaviour in its social and cultural aspects and thus its study must reflect the heterogeneity of these branches, as well as their common origin (human behaviour), and apply to it.

Therefore, sustainability is a key term both from and for a number of social science branches (economics, law, political science, communications and media, environmental studies) and, in each of them, has been perceived and employed differently across different eras. Since social science was gradually institutionalized

R. MacGregor Pelikánová, *Sustainability in Europe*,
https://doi.org/10.1007/978-3-031-74689-5_3

in the eighteenth and nineteenth centuries to become more systematic and global,[1] its branches followed these trends and became more organized and methodologically set.

The discipline of economics studies the production, distribution, and consumption of goods and services and sustainability is typically perceived as a viable manner to go ahead with such production, distribution and consumption.

Legal science, as known as law studies, targets the law by exploring legal systems as frameworks in which to interpret and use the law and sustainability. In law studies, sustainability is typically perceived as a suggestion or a command, hence its regime oscillates between rather facultative accountability, partially (morally) binding responsibility and mandatory liability. In this perspective, it is important to emphasize that *to count* means to measure and quantify and to be accountable is to explain what one has done with resources[2] and thus accountability has an individual, subject-oriented dimension not fully in the reach of the objective law.[3] Responsibility, as such, has Latin roots, see *respondere*, and means that someone has to answer for the effects caused by him or somebody else to an authority, and this authority evaluates its damages.[4] If this mechanism is incorporated into the legal system and this authority is a judge, we deal with a particular type of responsibility called liability, which is at the very heart of law studies.[5] In sum, accountability is basically about consciousness vis-à-vis to yourself, while responsibility is about moral duties to others (moral authorities, see ethics) and liability is about legal duties to the entire society (legal authorities, see law). Consequently, law studies and legal science are perhaps more preoccupied with the delineation between these three, rather than being concerned with the very definition and conceptual determination of sustainability. The target is the regime, not so much the term and its concept.

Political science studies systems of governance and power and the analyses of political activities, political thought, political behaviour, and associated constitutions and laws. So, for political science, sustainability is typically perceived as a compromise to be employed while governing in order to maintain stability, continuity, and effectiveness over time.

Communications and media studies information and its transfer and publication. Hence, sustainability is, for them, an information factor, value or determinant. From a rather neutral and informative term about the manner of management of natural (re)sources, sustainability became both a fashionable tool and a target of strategic management, with an unfortunately underplayed recognition of the importance of time, see the issue of short-termism.[6]

[1]Tóthová (2023), pp. 78–88.

[2]Stolowy and Ding (2017), p. 10.

[3]MacGregor Pelikánová (2023), p. 333.

[4]Schüz (2012), pp. 7–15.

[5]MacGregor Pelikánová and MacGregor (2020), pp. 74–108.

[6]Bansal and DesJardine (2014), pp. 70–78.

Environmental studies deal predominantly only with one pillar of sustainability, while history addresses sustainability with a minimal focus on this environmental pillar.

However, two branches, aka fields, of social science have a massive and, at the same time, deeply overlooked impact on sustainability and its employment—namely, linguistics and philosophy.

3.2 Sustainability in Linguistics

Linguistics is the scientific study of language and its structure. Traditionally, it entails the morphology (word structure analysis), syntax (sentence structure analysis, aka rules governing the structure of sentences), semantics (meaning of single and composed terms), phonetics (speech sounds and equivalent gestures in sign languages), phonology (the abstract sound system of a particular language), and pragmatics (how social context contributes to the meaning). Regarding the conceptual and terminological background of sustainability, the morphology, semantics and pragmatism are all highly relevant.

The morphology, combined with the etymology, of the term sustainability is obvious, because this word is a good example of a not purely Anglo-Saxon composition. Namely, similar to other legal English terms, it is the result of the history and political-social processes in which were involved the Anglo-Saxon, Jute and Frisian tribes, Latin-speaking missionaries, Scandinavian warriors and Norman conquerors. The term sustainability started to emerge in English documents in the sixteenth and seventeenth centuries. During this same era, French, as the language of the elites, was replaced by English in the fifteenth century, and Latin, as the language of the law and court proceedings, was replaced by English by the Proceedings in Courts of Justice Act in 1730.[7] The Latin translation of the verb *to sustain* is *retineo* or *sustento* and the French translation is *maintenir* or *soutenir*. In Choirs or Orchestras rehearsals even today, the Director frequently will use the Italian term *sostenuto*, meaning to sustain or hold a note beyond its full value. Thus, morphologically is at the very heart of this term, the holding action, i.e. to hold either by hand or otherwise in a reliable manner. In contrast, the German term, *Nachhaltigkeit*, has a stronger version of this *sus,* i.e. it is *nach* which means *after.* Indeed, the increased demand to make this concept long-term and even eternal led to a partial move in French from *soutenabilité* to *durabilité.* This clearly shows that Germans appreciated the importance of the durable and ongoing feature much earlier than the English and French and so the German *Nachhaltigkeit* is morphologically more obviously oriented towards eternity than the English *sustainability* or French *soutenabilité.* It can even be argued that the Latin-French roots of *sustainability* and the massive agricultural focus, as opposed to the German Hansa business focus, has, until recently,

[7] Schneiderová (2018), pp. 117–126.

linguistically undermined the need for the solidarity and co-operation between various generations. Therefore, the English morphology of *sustainability* is far from the currently expected *durability* or *permanency*.

The field of semantics studies the meaning in language of both single and composed terms and considers a number of universal approaches and rules (literal, etc.), as well as special branches' approaches and rules (golden, mischief and purposive), and even regionally particular approaches and rules (teleological due to the CJ EU determination to go for the spirit of the EU law via an autonomous pro-integration interpretation). Naturally, the semantic analysis of sustainability is not detached from other linguistic levels and it builds upon the above-mentioned *sustainability* word structure analysis (morphology) and the sentence structure analysis (syntax), while the context appreciation is critical. It cannot be overemphasized that the current EU language is an artificial construct on the supranational level to be employed nationally. Therefore, the meaning of *sustainability* is shaped by both EU demands and each EU member state's national realities.[8] Plainly, the alleged meaning assigned to sustainability, as well as to CSR, perhaps even CSV, by the EU policy- and law-making institutions, needs to be communicated to national authorities which will transpose it and apply it.[9] Naturally, both the EU and each EU member state's interpretation authorities, especially courts and judges, will consider a number of universal approaches and rules (literal, etc.), as well as special branches approaches and rules (golden, mischief and purposive), and even regionally particular approaches and rules (teleological). Undoubtedly, due to the consistent case law of the CJ EU, and its steady determination to go for the spirit of the EU law via an autonomous pro-integration interpretation, the teleological interpretation of sustainability and its regime is to prevail. It should be pointed out that the ultimate decision maker about the semantics meaning is the judge who, within the law dispute, does not start, but only arrives at a true meaning at its conclusion.[10]

Pragmatics is about the appreciation about how the social context contributes to the meaning and undoubtedly pushes the meaning of sustainability in the current EU even further away from the plain morphology analysis of Latin v French v German v English, and from the teleological semantics neologism bridging between supranational and national. In Europe, sustainability was always an abstract, heterogenous and conceptually unsettled argumentative command with at least partially obscure roots. Therefore, the social context has allowed variable interpretations going far above and beyond the teleological approach. Language is executed in concrete communicative acts and so the pragmatism points out how sustainability is shaped by interdisciplinary and two-ways communications.[11] Therefore, the evolutionary milestones representing contextual hallmarks shaping the very meaning of

[8] Ioriatti (2022), pp. 275–291.

[9] MacGregor Pelikánová and Sani (2023), pp. 813–851.

[10] Stamper (1991), pp. 219–244.

[11] Glogar (2023), pp. 1081–1107.

sustainability can be personalised. The advanced smart, sustainable and inclusive growth era of Jose Manuel Barroso and Jean-Claude Juncker with the Europe 2020[12] has been replaced, as of 1 December 2019, by an even more progressive pro-sustainable development era of Ursula von der Leyen and her Political Guidelines with six priorities,[13] in particular the European Green Deal,[14] with its aim to transform the EU's economy and society to *reach enhanced environmental sustainability.*[15] Although some critics perceive the EGD as a moderate organic continuation of pre-existing pro-sustainability trends slightly greening capitalism,[16] a strong tenor argues that it is not merely old wine in new wineskins, but instead a major societal transformation of the EU and that the time has come for greater sustainability.[17] As a matter of fact, the original version of the EGD was perceived by several influential streams as excessive and ultimately calls for more pragmatism, competitiveness and a revival of the economic pillar led to the Clean Industrial Deal, see the differences between Political Guidelines for 2019–2024[18] and Statement for 2024–2029.[19]

3.3 Sustainability in Philosophy

The term sustainability is clearly within the reach of another social science branch—philosophy. Of the three areas of philosophy—metaphysics (what is real?), epistemology (how do we know?) and axiology (what is right?), what is relevant are ontology from metaphysics (what is real and exists?) and ethics from axiology (what is (morally) right).

---

[12] COM(2010) 2020 Final Communication from the Commission—EUROPE 2020 A strategy for smart, sustainable and inclusive growth from 3 March 2010 ("Europe 2020").

[13] Von der Leyen U (2019) Political Guidelines for the next European Commission 2019–2024: A Union that strives for more. My agenda for Europe. Available from: https://www.europarl.europa.eu/resources/library/media/20190716RES57231/20190716R ES57231.pdf ("Political Guidelines").

[14] COM/2019/640 final Communication from the Commission to the European Parliament, the European Council, the Council, the European Economic and Social Committee and the Committee of the Regions: The European Green Deal from 11 December 2019 ("European Green Deal").

[15] Schunz (2022), p. 2085121.

[16] Ossewaarde and Ossewaarde-Lowtoo (2020), p. 9825.

[17] Tendero (2022), pp. 156–184.

[18] Von der Leyen U (2019) Political Guidelines for the next European Commission 2019–2024: A Union that strives for more. My agenda for Europe. Available from: https://www.europarl.europa.eu/resources/library/media/20190716RES57231/20190716RES57231.pdf("Political Guidelines").

[19] Von der Leyen U (2024) Statement at the European Parliament Plenary by President Ursula von der Leyen, candidate for a second mandate 2024–2029. Our first priority will be prosperity and competitiveness from 18 July 2024. Available from: https://neighbourhood-enlargement.ec.europa.eu/news/statement-european-parliament-plenary-president-ursula-von-der-leyen-candidate-second-mandate-2024-2024-07-18_en("Statement for 2024–2029").

Regarding ontology, it is obvious that sustainability and related concerns are real and they exist. The fact that the issue of sustainability is not illusory but real was a lesson learned by ancient agricultural societies, living near rivers, in a rather radical manner. Over succeeding centuries, this local and temporary concept in regard to agriculture has turned into a global and permanent concept reaching all fields. Even more specifically, since the twentieth century, the UN has kept tying the sustainability reality to advanced global solidarity goals and to fight against poverty worldwide, while the EU keeps tying the sustainability reality to advanced and systematic global environmental concerns.

Since the ontological question what is reality is both the alpha and omega of any discourse about the ephemeral term sustainability, it can be argued that the current metaphysical hesitation between these two sustainability pillars (social and environmental) is further carried by the axiology looking for the right choice, in particular ethics preferences. Indeed, sustainability was a much easier subject matter for philosophy two thousand years ago than now. The current concurrence, and often tension, between three pillars in the global setting leads rather to trade-offs than to a win-win result. The complexity of this issue became even more magnified by the assignment of the responsibility, if not liability, to nations and even individuals which are expected, if not directly required, to work towards the economic, environmental and social sustainability. Logically, this leads to the justification of such sustainability and such responsibility promoting regimes.

Thusly, it can be argued that the justification of sustainability is a task for axiology ethics, and the legitimization of sustainability a task for natural law and legal positivism theories. Observing the compromises of the Hobbes' social contract, sustainability can hardly be imposed by international or regional organizations, and the pathway to such responsibility and liability is highly questionable, while the concession of the utilitarian Bentham's pragmatism might suggest a trade-off sustainability and open doors to such a pathway. The deontological principles of Kant's judgment might make the choice of such a pathway extremely complex, because a categorical imperative, aka command, to be followed is definitely not obvious regarding sustainability in the twenty-first century. Indeed, even the universalism of Aristotle's law of nature leads to the complex setting of a general complete justice, aka rightness, and particular justice. Namely, the particular justice is either public distributive (appreciative with rewards or vindictive with punishments) or private corrective (restitutive for involuntary behaviour or commutative for voluntary contracts). The public distributive justice goes for a diagonal positive conjunction and means that authority gives to subjects, as opposed to private corrective justice when the authority monitors re-distribution between subjects. This Aristotle-like scheme shows the clarity about historical sustainability and the confusion about current sustainability, which mixes distributive justice (an authority imposing and paying) and corrective justice (an authority moving resources from one to another) and where such an authority is even questionable. Unsurprisingly, the justification of modern sustainability might appear shaky and thusly insufficient to pass the legitimacy test to further push responsibility to liability.

The theories of law and legal justification have dominated legal philosophy in Europe—the natural law (what is accepted as morally binding) and legal positivism (what is binding based on power). The natural law ties together moral merits with enforceable standards and refers to the ultimate authority of someone other than a human being (e.g. God). However, in the context of consequentialism, in particular utilitarianism by John Austin, the pendulum kept moving from "God made law" of natural law to Man- made law of legal positivism by reaching the peak with the famous Hart's Concept of Law in 1961.[20] In the view of such a legal positivism, rules are made by humans and that there is no inherent or necessary connection to morality. Hence, once an authority defines the sustainability and its regime, no further justification or replies to questions are needed. However, the pendulum swung back in 1986, due to Dworkin's Theory of law as integrity, stating that judges interpret the law in terms of consistent moral principles, especially justice and fairness. This revives the question about which sustainability definitions and regimes are just and fair and, in particular, which pro-sustainability duty can be justly and fairly imposed. The following year, the Brundtland report answered it in a rather radical manner, by demanding large financial spending "in the short term in such fields as renewable energy development, pollution control equipment, and integrated rural development. Developing countries will need massive assistance for this purpose, and more generally to reduce poverty. Responding to this financial need will be a collective investment in the future" (6.94 Brundtland Report). The pre-cursor of the EU, the European Economic Community, was explicitly mentioned as one of the sources and channels for such support (6.101 Brundtland Report). Undoubtedly, the EU Commission accepted this, see Europe 2020 and Political Guidelines, but Europeans keep posing questions about the *sustainability* and its regime and their justifications. These questions cannot be answered without a consideration of subjective ethical features reflecting one's personal morals. Indeed, since the great challenge of sustainability is to overcome personal selfishness[21] and the ultimate drive should be for respect and against waste,[22] the philosophical dimension of the discussion about sustainability cannot be overemphasized.

3.4 Meta-Sustainability

Conceptually and terminologically, sustainability is an ephemeral, evolving, rather vague, and strongly contextually dependant, term,[23] which arguably always targets the ongoing survival of humans, and this either locally (early approaches) or

[20] Payne (1978), pp. 11–17.

[21] D'Adamo et al. (2024), p. 5049.

[22] MacGregor Pelikánová et al. (2024), pp. 1–29.

[23] White (2013), pp. 213–217.

regionally and nationally (newer approaches) or globally (the current approach).[24] Inherently, the abstract concept of sustainability conflicts with the eternal human self-control problem leading to the inclination for immediate gratification and self-indulgence in a way humans do not appreciate but still do.[25] At the same time, sustainability and sustainable development are EU core principles, firmly anchored in the EU Treaties[26] and both vigorously and rigorously projected in EU secondary law as well as EU policies and strategies for over one decade.[27] As a matter of fact, the concept of sustainability has been shaping the EU law and policies and, vice versa, the EU and policies have shaped the concept of sustainability, but recently even the core concepts and principles from rather other fields are transformed due to sustainability, see e.g. the dramatic change of consumer as *homo economicus*.[28]

The concept of pragmatic sustainability may be contrasted with that of ideological sustainability.[29] Namely, on one hand, sustainability can be perceived either as merely about finding a compromise, leading to set goals, such as climate neutrality 2050, or as a complete shift to a solidarity brotherhood of humans global-wise[30] with a high respect for nature.[31] Sustainability was and remains a multi-disciplinary challenge calling for functionalist, analytical, contextual and even historical comparison.[32]

Since its emergence a millennium ago, sustainability has been intimately linked to both the pragmatic product generating economics and to the philosophical search for justice and fairness. The society was to be provided for and this providing was to be done rightly. Hence, the original balancing was between agricultural economics and social justice with a very small, if any, impact of other social science branches, such as law or environmental protection. This original dichotomy was addressed locally and in an authoritarian manner by following distributive and not corrective justice. This is definitely not the case in the twenty-first century.

Indeed, sustainability in the twenty-first century is globally omnipresent, rests on three pillars and demands actions of states as well as individuals. However, the aforementioned global dimension brings about the interaction of different cultures and civilizations sharing different traditions, sustainability perceptions and duty imposition readiness. Europe, with its background of an orientation on individual responsibility, keeps moving towards a pro-sustainability regime with legal duties in and even out of its jurisdiction. The European Commission offers a battery of justifications to do so without worrying about any definition and concept ambiguity.

---

[24] Brown et al. (1987), pp. 713–719.

[25] O'Donoghue and Rabin (2000), pp. 233–250.

[26] Borchardt et al. (2022).

[27] Balcerzak et al. (2023), pp. 199–212.

[28] MacGregor Pelikánová (2024), pp. 218–235.

[29] D'Adamo et al. (2024), p. 5049.

[30] D'Adamo and Gastaldi (2023), p. 14094.

[31] Dodman et al. (2020), pp. 3–10.

[32] Tóthová (2023), pp. 78–88.

This trend of progressive integration of the rather vaguely defined concept of sustainability, with the ephemeral conjunction of sustainable development, in the international and EU policies and law, has been observed and heavily discussed by academic literature from the era of Europe 2020[33] as well as today.[34] Although the conceptual and terminological ambiguity has been repeatedly confirmed, the prevailing literature trend attempts to create a connection to CSR, CSV and even more advanced social concerns.[35] This often leads to the question of the conflicting prevalence (typically juxtaposing environmental and social concerns) and even the feasibility (typically juxtaposing economic concerns to environmental and social concerns).[36] A project entailing the extraction and analysis of over one hundred definitions of sustainability in the current European academic literature revealed that sustainability remains an elusive concept which means different things to different people and thus it is difficult to define.[37] It is noteworthy that the terms most frequently used in the definition of sustainability are environment, social, economic, life, system, nature, resources, human and development, while growth and equity are significantly less frequently included. In addition, decisions about the concept of sustainability and its application are done by various policy—and law-makers and agents, but, interestingly, there is a clear central driver—the trust,[38] which is linked to two core values—no waste and respect.[39]

Indeed, the term *sustainability* is manipulated and misunderstood. In addition, the term *sustainable development* represents an inherent linguistic contradiction, see the explanation by Daly about the growing economic subsystem which incorporates an increasing proportion of the total ecosystem.[40] Arguably, sustainability is unsustainable and ultimately doomed due to its lack of pragmatism, shortcuts ignoring not only the business but, as well, technical reality,[41] disinterest for compromising and balancing, disrespect of historical lessons, patronizing political correctness, and, ultimately, its incapacity to inspire the very needed support across the entire society via a multi-stakeholder model. This discourse has been taking place on various platforms in an interdisciplinary manner, but has not changed the course of action of the UN and EU. The UN keeps understanding sustainable development as a conjunction of sustainability perceived as durability and of

---

[33] Van Rijswick (2012), pp. 1–6.

[34] Menghwar and Daood (2021), pp. 466–485.

[35] Kramer and Pfitzer (2016), pp. 80–89.

[36] MacGregor Pelikánová et al. (2021), p. 3715.

[37] White (2013), pp. 213–217.

[38] Eger et al. (2019), pp. 141–153.

[39] MacGregor Pelikánová et al. (2024), pp. 1–29.

[40] Daly (1991), p. 39—"*In its physical aspects, the economy is an open subsystem of the terrestrial ecosystem, which is finite in material resources. As the economic subsystem grows, it incorporates an increasing proportion of the total ecosystem. For this reason, development is not sustainable. The term sustainable development, as it applies to the economy, is a contradiction....*"

[41] D'Adamo et al. (2023), pp. 840–852.

development, perceived as extensiveness,[42] and keeps turning it into a global command resting on three pillars: environmental (planet), social (people) and economic (profit). This is implied by the UN Brundtland Report, made actionable by the UN Agenda 2030 in 2015,[43] connected to the move from CSR to CSV[44] and fully endorsed by the EU. The wording of the European Action for Sustainability from 2016[45] is self-explanatory: "Sustainable development has since long been at the heart of the European project. The EU Treaties give recognition to its economic, social and environmental dimensions which should be addressed together. The EU is committed to development that meets the needs of the present without compromising the ability of future generations to meet their own needs. ... European societies today face many sustainability challenges from youth unemployment to ageing populations, climate change, pollution, sustainable energy and migration. We must tackle current challenges and prepare for the future, responding to the pace and complexity of global change and the demands of an increasing global population." Hence the EU keeps combining sustainability and sustainable development and follows the UN vision of the global and intergenerational solidarity with all 17 SDGs. This is pushed even further by the Political Guidelines of Ursula von der Leyen, see their first section 1.A European Green Deal with her strong subtitle quote "I believe that what is good for our planet must be good for our people, our regions and our economy. We will ensure a just transition for all" on page 6. It cannot be overemphasized that the Political Guidelines include six priorities of which the first one is the European Green Deal, while the New Political Guidelines continues it with the new European Prosperity Plan and a Clean Industrial Deal. Who would have imagined that all this followed from an autocratic management of floods in local agriculture? Who would imagine that the currently omnipresent and so heavily discussed concept of sustainability entails many aspects which have not yet received an in-depth exploration to be properly understood?[46]

Legal Documents

COM(2010) 2020 Final Communication from the Commission – EUROPE 2020 A strategy for smart, sustainable and inclusive growth from 3 March 2010 ("Europe 2020")

---

[42] Duran et al. (2015), p. 806–811.

[43] Van Tulder and Keen (2018), pp. 315–332.

[44] Porter and Kramer (2011), pp. 62–77.

[45] COM(2016) 739 Final Communication from the Commission to the European Parliament, the Council, the European Economic and Social Committee and the Committee of the Regions: Next steps for a sustainable European future - European action for sustainability. SWD(2016) 390 final from 22 November 2016 ("European Action for Sustainability").

[46] D'Adamo et al. (2024), p. 5049. https://doi.org/10.3390/su16125049.

COM(2016) 739 Final Communication from the Commission to the European Parliament, the Council, the European Economic and Social Committee and the Committee of the Regions: Next steps for a sustainable European future - European action for sustainability. SWD(2016) 390 final from 22 November 2016 ("European Action for Sustainability")

Von der Leyen U (2019) Political Guidelines for the next European Commission 2019–2024: A Union that strives for more. My agenda for Europe. Available from: https://www.europarl.europa.eu/resources/library/media/20190716RES57231/20190716R ES57231.pdf ("Political Guidelines")

COM/2019/640 final Communication from the Commission to the European Parliament, the European Council, the Council, the European Economic and Social Committee and the Committee of the Regions: The European Green Deal from 11 December 2019 ("European Green Deal")

Von der Leyen U (2024) Statement at the European Parliament Plenary by President Ursula von der Leyen, candidate for a second mandate 2024-2029. Our first priority will be prosperity and competitiveness from 18 July 2024. Available from: https://neighbourhood-enlargement.ec.europa.eu/news/statement-european-parliament-plenary-president-ursula-von-der-leyen-candidate-second-mandate-2024-2024-07-18_en ("Statement for 2024-2029")

References

Balcerzak A, MacGregor RK, MacGregor Pelikánová R, Rogalska E, Szostek D (2023) The EU regulation of sustainable investment: the end of sustainability trade-offs? Entrep Bus Econ Rev 11(1):199–212. https://doi.org/10.15678/EBER.2023.110111

Bansal S, DesJardine MR (2014) Business sustainability: it is about time. Strateg Organ 12(1): 70–78. https://doi.org/10.1177/1476127013520265

Borchardt S, Barbero Vignola G, Buscaglia D, Maroni M, Marelli L (2022) Mapping EU policies with the 2030 agenda and SDGs, EUR 31347 EN. Publications Office of the European Union, Luxembourg., ISBN 978-92-76-60474-7. JRC130904. https://doi.org/10.2760/87754

Brown BJ, Hanson ME, Liverman DM, Merideth RW (1987) Global sustainability: toward definition. Environ Manag 11:713–719. https://doi.org/10.1007/BF01867238

D'Adamo I, Gastaldi M (2023) Monitoring performance of sustainable development goals in the Italian regions. Sustain For 15:14094. https://doi.org/10.3390/su151914094

D'Adamo I, Di Carlo C, Gastaldi M, Rossi EN, Uricchio AF (2024) Economic performance, environmental protection and social progress: a cluster analysis comparison towards sustainable development. Sustainability 16:5049. https://doi.org/10.3390/su16125049

D'Adamo I, Gastaldi M, Ozturk I (2023) The sustainable development of mobility in the green transition: renewable energy, local industrial chain, and battery recycling. Sustain Dev 31(2): 840–852. https://doi.org/10.1002/sd.2424

Daly HE (1991) Sustainable development: a contradiction. Desarro Base 15(3):39

Dodman M, Aillon JL, Arrobbio O, Camino E, Colucci-Gray L, Ferrara E, Folco S (2020) To connect or not to connect. Is that the question? Vis Sustain 13:3–10. https://doi.org/10.13135/2384-8677/4602

Duran DC, Gogan LM, Artene AE, Duran V (2015) The components of sustainable development - a possible approach. Proc Econ Finance 26:806–811. https://doi.org/10.1016/S2212-5671(15)00849-7

Eger C, Miller G, Scarles C (2019) Corporate philanthropy through the lens of ethical subjectivity. J Bus Ethics 156(1):141–153. https://doi.org/10.1007/s10551-017-3551-1
Glogar O (2023) The concept of legal language: what makes legal language 'legal'? Int J Semiot Law 36:1081–1107. https://doi.org/10.1007/s11196-023-10010-5
Ioriatti E (2022) Common contexts of meaning in the European legal setting: opening Pandora's box? Int J Semiot Law 36:275–291. https://doi.org/10.1007/s11196-022-09919-0
Kramer MR, Pfitzer MW (2016) The ecosystem of shared value. Harv Bus Rev 94:80–89
MacGregor Pelikánová R (2023) Accounting science's historical development and its role in social science & accounting theories and their application in accounting sciences. In: Knihová L (ed) Applying interdisciplinarity to globalization, management, marketing, and accountancy science. IGI Global, p 333. https://doi.org/10.4018/978-1-7998-8669-3.ch009
MacGregor Pelikánová R (2024) EU legislative proposals to involve customers in the greening of the economy. Lawyer Quart 2:218–235. https://tlq.ilaw.cas.cz/index.php/tlq/article/view/597
MacGregor Pelikánová R, MacGregor RK (2020) The EU puzzling CSR regime and the confused perception by ambassadors of luxury fashion businesses: a case study from Pařížská. Cent Eur Bus Rev 9(3):74–108. https://doi.org/10.18267/j.cebr.240
MacGregor Pelikánová R, Sani M (2023) Luxury, slow and fast fashion—a case study on the (un)sustainable creating of shared values. Equilibrium Quart J Econ Econ Policy 18(3): 813–851. https://doi.org/10.24136/eq.2023.026
MacGregor Pelikánová R, Němečková T, MacGregor RK (2021) CSR statements in international and Czech luxury fashion industry at the onset and during the COVID-19 pandemic—slowing down the fast fashion business? Sustainability 13(7):3715. https://doi.org/10.3390/su13073715
MacGregor Pelikánová R, Sani M, Rubáček F (2024) Sustainable and responsible creation of shared values in the fast fashion industry—I will believe it when I see it. Vis Sustain 21(8873):1–29. https://doi.org/10.13135/2384-8677/8873
Menghwar PS, Daood A (2021) Creating shared value: a systematic review, synthesis and integrative perspective. Int J Manage Rev 23(4):466–485. https://doi.org/10.1111/ijmr.12252
O'Donoghue T, Rabin (2000) The economics of immediate gratification. J Behav Decis Mak 13(2): 233–250
Ossewaarde M, Ossewaarde-Lowtoo R (2020) The EU's green deal: a third alternative to green growth and degrowth? Sustainability 12:9825. https://doi.org/10.3390/su12239825
Payne M (1978) The basis of law in Hart's "the concept of law.". Southwest J Philos 9(1):11–17. https://www.jstor.org/stable/43155198
Porter ME, Kramer MR (2011) Creating shared value. How to reinvent capitalism—and unleash a wave of innovation and growth. Harv Bus Rev 89(1–2):62–77
Schneiderová A (2018) Historical background to English legal language. J Mod Sci 2(37):117–126. https://doi.org/10.13166/jms/92522
Schunz S (2022) The 'European Green Deal'—a paradigm shift? Transformations in the European Union's sustainability meta-discourse. Polit Res Exchange 4(1):2085121. https://doi.org/10.1080/2474736X.2022.2085121
Schüz M (2012) Sustainable corporate responsibility—the foundation of successful business in the new millennium. Cent Eur Bus Rev 1(2):7–15
Stamper RK (1991) The role of semantics in legal expert systems and legal reasoning. Ratio Juris 4(2):219–244
Stolowy H, Ding Y (2017) Financial accounting and reporting : a global perspective, 5th edn. Cengage Learning
Tendero M (2022) The time has come! The development of the European environmental conscience: evidence from the eurobarometer surveys from 1974 to 2020. In: Hoerber T, Weber G (eds) The European environmental conscience in EU politics. Routledge, Abingdon, pp 156–184
Tóthová M (2023) Several aspects of the comparative method in law. Lawyer Quart 13(1):78–88

Van Rijswick HFMW (2012) The road to sustainability: how environmental law can deal with complexity and flexibility. Utrecht Law Rev 8(3):1–6. https://utrechtlawreview.org/articles/10.18352/ulr.202

Van Tulder R, Keen N (2018) Capturing collaborative challenges: designing complexity-sensitive theories of change for cross-sector partnership. J Bus Ethics 150:315–332. https://doi.org/10.1007/s10551-018-3857-7

Von der Leyen U (2019) Political Guidelines for the next European Commission 2019–2024: A Union that strives for more. My agenda for Europe. Available from: https://www.europarl.europa.eu/resources/library/media/20190716RES57231/20190716RES57231.pdf. Accessed 18 Nov 2023

White MA (2013) Sustainability: I know it when I see it. Ecol Econ 86:213–217. https://doi.org/10.1016/j.ecolecon.2012.12.020

Chapter 4
Origins of the Concept of Sustainability

The concept of sustainability has millennial roots which reflect predominantly the continental law tradition. The concept of sustainability has always mirrored value judgments about justice in distributing and using resources[1] and was reflected by both secular administrations, see the Mesopotamian and Egyptian flooding and cultivation management, as well as by religious teachings, see Bible verses about the 7 years of abundance followed by 7 years of shortage and famine.[2] It is tied to Aristotle's teaching distinguishing various types of justice and endorsing a pro-merit approach based on virtues. For a millennium, these ideas of Aristotle have been providing the general direction for both the present and the future of societies on the European continent. The implied multi-spectral commitment to sustainability and long-term preferences over immediate consumption gratification was one of the key factors for the hundreds of years' long glory of the Ancient Roman Empire. The erosion of this economic, environmental and social commitment contributed to the fall of both the Western, and later, the Eastern Roman Empire. The renaissance of this commitment, along with the exit from the chaos allowed to enter in the new era of sustainability which ultimately leads to our current time. Therefore, any attempt at understanding the roots of (current) sustainability needs to go back to antiquity and examine the prevailing philosophical dimension of sustainability (Sect. 4.1) and the social and economic setting relevant for sustainability in Antiquity (Sect. 4.2).

[1] MacGregor Pelikánová and MacGregor (2020), pp. 74–108.

[2] MacGregor Pelikánová and Hála (2021), p. 390.

R. MacGregor Pelikánová, *Sustainability in Europe*,
https://doi.org/10.1007/978-3-031-74689-5_4

4.1 The Philosophical Dimension of Sustainability in Antiquity: Sustainability as a Virtue

Since sustainability is about understanding and managing reality, it is clearly a subject matter of metaphysics. In antiquity, the starting point of a philosophical perception of sustainability consisted of the metaphysical analysis of very tangible and practical re-occurring issues. Almost all of them were directly or indirectly linked to agriculture, the production feeding the population. Neither special logical arguments nor axiology justification were needed, the philosophical setting closely mirrored the pragmatism of the survival instinct of three thousand years ago. The epistemology of Plato and the fluid distinction between the tangible and intangible, along with the trust in the knowledge and consciousness inevitably led to intrapersonal perceptions and intrapersonal justice. However, this way of thinking was not practical in the situation of flooding and natural disasters as well as social unpredictability. Indeed, civilizations and kingdoms prior to the Greek municipal states could not prosper without a sophisticated central organization under the auspices of autocratic leaders, their elites. Mesopotamia, Babylon, Egypt, etc. developed advanced social and economic systems relying on co-ordinated actions and using human power without a juridical personality, in short, slaves. The evil was chaos and had to be fought on all levels—by building pyramids, by creating complex bureaucratic structures and by managing the (agricultural) production sustainably, by developing schools of philosophy addressing the study (epistemology) of the reality (metaphysics) and the justice (axiology).

Philosophy, as the love of wisdom via a systematic study of general and fundamental questions concerning our being, emerged in Europe in the sixth century BC. Socrates, Plato and Aristotle[3], as well as other philosophers preceding them, laid down concepts and doctrines recognized even today.[4]

Considering the needs of the ancient autocratic and centralized societies, Socrates and Plato's abstract ideas about justice as a balance of reason, spirit, and appetite appeared abstract to the pragmatic state, economics and law settings. In contrast, Aristotle's perception of reality and justice matched it visibly. Aristotle promoted metaphysics and combined the substance (form) and essence (matter) and, due to his realism, if not pragmatism, his epistemology is based on the observation and study of existing things following inductive thinking. Therefore, Aristotle realized that, regarding certain aspects, individuals are/should be considered equal and get equal shares while, regarding other aspects, individuals are unequal and should receive unequal shares. Therefore, Aristotle followed Plato's idea about people separated by differences in their capacities and went further by conducting an analysis of justice between free men.[5] The observation of the reality of over two thousand years ago

[3]Blackson (2011).

[4]Copleston (2003).

[5]Johnston (2011).

from Aristotle's perspective reveals that sustainability is about addressing the needs of many subjects by using many resources. The axiology questions about the legitimate, effective and efficient materialization of these commands are overshadowed by the pragmatic revelation that co-ordinated behaviours are needed. Plainly, sustainability, advancing actions and omissions, must entail various subjects dealing with various resources in a continual, viable and feasible manner. This means that those persons who are these subjects have to share a general agreement about their sustainability preferences and about how these preferences are to be achieved justly and fairly. Naturally, each person has his or her own unique personal moral principles that govern his or her particular behaviour and preferences. Since these individual moral principles are not very easily able to be reconciled and consolidated, a systematization of the concept of sustainability by ethics is extremely challenging and moral philosophy has been pre-occupied by it while offering various approaches.

Nevertheless, it is commonly accepted that such a distribution and consumption is tied to Aristotle's teaching distinguishing general (complete) justice, similar to our rightness and perhaps even the rule of law, and particular justice, which can be either public distributive or private corrective. Public distributive justice entails a diagonal conjunction and means that one authority makes a distribution to a member of society in order to reward them (*appreciative justice*) or to punish them (*vindicative justice*). The private corrective justice (*rectifactory justice)* entails an average formation and the authority monitors the redistribution and transfer between members of the society which provides each other a compensation, either in the case of an involuntary interaction (*restitutive justice)* or in the case of a voluntary contract (*commutative justice*).

Inherently, public distributive justice is basically about the socially just allocation of resources, goods and opportunities in society, and such a distribution is to be done based on a distributive criterion. Aristotle distinguishes between the arithmetical flat equality in which all persons should receive equal shares regardless of their merits (everybody gets the same equal share) and a geometrical equality where unequal persons receive unequal shares due to proportionate equality reflecting merits based on the contribution principle. The underlying idea of Aristotle is a pro-merit approach, which means that the greatest privileges go to those male aristocrats who exhibit the greatest virtue rather than to those who have the greatest wealth, the greatest military strength, or the most friends. For over two millennia, these ideas of Aristotle's have been providing the general direction for the future and they influence the concept of sustainability.[6] This was done in the context of Christianity and in particular the teachings of Thomas Aquinas, holding Aristotle in the highest esteem, which complemented the teachings of Saint Augustine, inspired by Plato.

The very strong call for, and defence of, sustainability is the red thread going through the Bible, both the Old Testament and New Testament, demonstrated via stories, songs, and parables set in the ancient society and thus they need to be

[6] Balcerzak and MacGregor Pelikánová (2020), p. 95.

contextually so understood. Regarding the Old Testament, it could be proposed that the story of the 7 years of great abundance (Genesis 41:29) and the 7 years of famine (Genesis 41:54) is an archetypical description of sustainability, its roots and implications. Regarding the New Testament, from the total of more than 40 parables told by Jesus, almost all address sustainability and almost each of them from a different angle and even their titles are self-explanatory—Wise and Foolish Builders (Matthew 7:24–27, Luke 6:47–49), Growing Seed (Mark 4:26–29), The Rich Fool (Luke 12:13–21, Unfruitful Fig Tree (Luke 13:6–9), The Weeds (Matthew 13:24–30, 36-4), The Seed (Matthew 13:31–32, Mark 4:26–30, Luke 13:18–19), The Lost Sheep (Matthew 18:12–14, Luke 15:3–7), The Talents (Matthew 25:14–30, Luke 19:11–27), The Sheep and the Goats (Matthew 25:31–46). Last, but not least, the seventh commandment prohibits business fraud, wilfully damaging private or public property, the private abuse of common goods, work deliberately done poorly and wasted.[7]

The connection between sustainability and Christianity, with the Bible and its call for stewardship of God's creation and responsibility for its care, has been academically and practically explored for hundreds of years,[8] while reflecting both on the social dimension[9] and environmental dimension.[10] Christianity was founded on the doctrine that humans have the individual capacity and responsibility to determine their own actions, that by these actions they should do good and that they are to be treated equally.[11]

John Evelyn has heavily referred to the Bible in his *Sylva or a Discourse of Forest Trees and the Propagation of Timber and His Majesties Dominions* (1662) and included long quotations by Greek and Roman authors, such as Pliny, Virgil and Ovid as well as more contemporary authorities, such as William Shakespeare.[12] Similarly Hans Carl von Carlowitz in his *Sylvicultura oeconomica* (1713) quoted from Genesis, his starting point was that a man does not live anymore in the garden of Eden and his pious Lutheran attitude endorsed the motto "*Bebauen und bewahren*", i.e. cultivate and preserve.[13]

Therefore, it can be safely stated that Sustainability European Style has roots linked to the metaphysical understanding of reality and axiological justice as proposed by Greek philosophers, in particular Aristotle, and emerged in the continental law setting of the growing agricultural-slave empires sharing the continental law traditions and was impacted by Christianity.[14] Sustainability styles in other regions might have similar roots, but still their particularisms are heavily engraved into them,

---

[7] US Conference of Catholic Bishops (2012), p. 146:508.

[8] Horrell et al. (2008), pp. 219–238.

[9] Tkáčová and Slivka (2021), p. 882. https://doi.org/10.3390/rel12100882.

[10] Abukabar (2025).

[11] Stark (2005), pp. 24–29.

[12] Grober (2007).

[13] Grober (2007).

[14] MacGregor Pelikánová et al. (2021), pp. 973–1009.

see the massive colonial and imperial impact on the development of not only environmental sustainability in Africa,[15] the influential Conservation Movement in the USA with the Forest Reserve Act of 1891[16] or the significance of the combination of a low population density, abundant natural resources, high labour productivity, political stability and developed international trade in Asia in the 18th and 19th centuries.[17] Indeed, the roots, evolution and development of sustainability have always been strongly dependent upon the pertinent very complex context.

4.2 Ancient Times: Sustainability as a Command

Ancient civilizations in Europe or near to Europe needed an enhanced central organization due to their natural and climate setting. Plainly, the rivers of the Euphrates, the Tigris, the Nile, etc. needed to be managed and a stable agricultural system nourishing the population had to be organized and maintained. Such a pragmatic, agricultural and rural- management oriented command was projected in the law, in fact the first organized collection of legal forms in a tangible format did this.

Indeed, the Babylonian Code of Hammurabi from 1755 – 1750 B.C.,[18] the oldest legal Code from the ancient Near East, includes almost all legal norms inserted between its massive prologue and epilogue.[19] These legal norms are expressed in the Babylonian dialect of Akkadian and written down in a version of Sumerian cuneiforms with pictograms reduced to a high level of abstraction. They demonstrate both strict punishment systems, amounting perhaps even to a hardship as well as aspects of morality, such as the demand for procedural justice by requiring evidence to be presented under the penalty of perjury (3–12) and setting rules for business relationships (100–126). The recording of transactions was done by a *scribe* on a moist clay tablet while parties added their „signatures" by impressing their respective seals.[20] These Babylonian *scribes* were well-respected professionals performing tasks nowadays done by notaries public, lawyers, accountants and auditors.

The underlying principle of the Code of Hammurabi was *lex talionis* (eye for an eye), subjects are not equal before the law, slavery is an integral part (278–282) and Hammurabi is called pious and the people's shepherd (Prologue) to organize and to take care of them. Naturally, the authority of Hammurabi was divine and beyond question. And beyond any question, as well, are his agricultural sustainability

---

[15] Kwashirai (2013), pp. 166–196.

[16] An Act to repeal timber-culture laws, and for other purposes, 16 U.S.C. ch. 2 from third March 1891 ("Forest Reserve Act of 1891").

[17] Bassino (2021), pp. 146–168.

[18] Code of Hammurabi, 1755 - 1750 B.C., a collection of 282 rules.

[19] MacGregor Pelikánová (2023), p. 333.

[20] Chiera (1938).

projects, such as the set of irrigation canals with the most important, *Nar Hammurabi,* to carry water from the new course of the Euphrates to the to the city of *Kish* in the direction of *Umma* and then to the city of *Larsa* to empty afterwards in the Gulf.[21]

Approximately one thousand years after the Code of Hammurabi, the Roman kingdom emerged. In 509 BC, it was replaced by the Roman Republic which promptly witnessed an increasing tension between patricians and plebeians which led to the promulgation of the The Law of the Twelve Tables (*lex duodecim tabularum*) in 449 BC.[22] The Laws of the Twelve Tables is recognized as a consolidation of earlier traditions and as the foundation of Roman law, i.e. the very roots of laws belonging to the continental law family, i.e. basically all national laws of the current EU member states. The majority of provisions addressed various rights and duties of Roman citizens and procedures for their enforcement and in particular included definitions and described in detail the regime of special and/or controversial institutions and concepts. Certain aspects of sustainability were probably beyond question and were not explicitly covered, but some aspects of sustainability were not obvious in the fifth century BC and so were included e.g., in Table VI and Table VIII.

Table VI. Ownership and Possession protected agricultural structures, see e.g., "One shall not take from the framework timber fixed in buildings or in vineyard … One shall be permitted neither to remove nor to claim stolen timber fixed in buildings or in vineyards, … but against the person who is convicted of having fixed such timber there an action for double damages shall be given" (Table VI.8).

Table VIII. Torts or Delicts enacts the option of compensation to avoid the crude versions of *lex talionis*, see "If anyone has broken another's limb there shall be retaliation in kind unless he compounds for compensation with him. If one commits an outrage against another the penalty shall be twenty-five asses" (Table VIII.2-4). In addition, there were provisions regarding agriculture and gardening, which made it into almost all Codes and belong to the universally recognized general practice, see "If fruit from your tree falls onto my farm and if I feed my flock off it by letting the flock onto it…. no action can lie against me either on the statute concerning pasturage of a flock, because it is not being pastured on your land, or on the statute concerning damage caused by an animal .." (Table VIII.7). Such production-collection rules, important for all pillars of sustainability, were complemented by financial-management rules with a similar effect, see "No person shall practice usury at a rate of more than one twelfth" (Table VIII.18a) and 20a. "If guardians are suspect in their administration there shall be the right to accuse them as such …" (Table VIII.20a) and "If … guardians steal a ward's property … there shall be an action … against a guardian for double damages; each guardian shall be held for the entire sum" (Table VIII.20b). Therefore, it can be argued that the local-regional agricultural focus was broadened and expanded into the sphere of general business

---

[21] Adamo and Al-Ansari (2020), pp. 87–111.

[22] Lex Duodecim Tabularum—Duodecim Tabulae (Law of the Twelve Tables), 449 BC.

management and asset management with responsibility, as a matter of fact for guardians jointly and severally liable.

The Roman Republic was replaced by the Roman Empire in 27 BC. However, during the second and third centuries AD, the Roman Empire experienced the autocratic divine governance, more and more bureaucratic administration, and a slightly increasing nepotism and self-indulgent style. This culminated in a set of serious crises threatening the very existence of the Roman Empire, which were overcome by the Emperor Constantine the Great. In lines of the iconic motto "Never waste a good crisis,"[23] he reunited the empire, won major military battles, founded the New Rome of Constantinople, and made monetary and administrative reforms. He mistrusted provincials and officials, especially if of the equestrian order (he had no use for senators nor members of established aristocratic families) and demanded transparency, free and timely access to justice and record keeping. Christianity played a leading role in the triumph of Constantine by providing him substantial and well-organized urban support[24] and by helping him to successfully unite the disintegrated Roman Empire, create the capital Constantinople, make monetary reforms, endorse Christianity and a new dimension of the ephemeral justice.[25] Constantine's Christianity united the constantly warring, divided Europe in 306–337 AD while relying on a set of pro-sustainability commands, such as waste avoidance, and propagation of justice,[26] efficacy of legislation with the leaning of bureaucracy.[27] The implied legal rules, about two hundred years later, were recorded in the *Corpus Iuris (or Juris) Civilis*[28] and, a few hundred years after that, found their way back to Europe, via Spanish academics, North Italian merchants and French managers.

The economic and agrarian models in ancient civilizations focused on longer time periods and relied heavily on long-term planning while rejecting immediate gratification from the exhaustion of available resources, and this was reflected by the law, see the Roman law as the foundation of the continental law tradition.[29] While Greek philosophy, along with Christianity, set conceptually the roots of sustainability, the Roman Empire managed to give this concept substance and materialized the creation of the roots of sustainability and its regime. Roman scholars and jurists have explained all this extensively. The recollection of them was incorporated in *Corpus Iuris Civilis* in the sixth century AD. They were rediscovered and subsequently taught in Bologna since the eleventh century. This transfer of legal knowledge was complemented by the transfer of knowledge about the philosophical teachings of

[23] Van Tulder and Van Mil (2023), pp. xxxi–xxxiii.

[24] Stark (2017), p. 57.

[25] MacGregor Pelikánová (2017), pp. 81–98.

[26] Dillon (2012), pp. 90–118.

[27] Dillon (2012), pp. 156–191.

[28] *Corpus Iuris Civilis (Codex, Digest or Pandects, Institutes/Institutiones + Novellae Constitutiones)*,529–534.

[29] MacGregor Pelikanova and Sani (2023), pp. 813–851.

Aristotle and other authorities, which was facilitated by the saving of the library collection of Augustine of Hippo (St. Augustine) in 430 from destruction by Vandals and its subsequent transport to Europe to be preserved in monasteries. This was the foundation for a new stage of evolution and the pre-cursor of many nowadays sustainability practices, see e.g. the Augustinian perspective on sustainability which encourages sharing and living simply[30] with our current circular and sharing economy drive.[31]

In short, Sustainability European Style was conceived thousands of years ago, prospered during the Roman Empire, became transferred to the early medieval era thanks to merchants bringing *Corpus Iuris Civil* and monks saving and disseminating prior academic and theoretical works, received a new and very practical multi-dimension by Franks on the border of modern France and Germany, to be advanced autonomously in Italy, England and France. Then it was further expanded by German admirers of Hansa and pupils of English and French experts, and later on moved to even higher levels and dimensions by the EU, especially under the presidency of Jean-Claude Juncker and Ursula von der Leyen. Arguably, scarcity is the mother of invention,[32] crises accelerated pre-existing trends[33] and the strength of European civilization is in its values recognizing individual responsibility and supporting creativity. Sustainability in Europe has to reflect it.

4.3 Different Approaches to Management of Resources

As already pointed out, the roots and concepts, as well as the evolution and the current meaning of sustainability are perceived dramatically on the global level,[34] on the level of the EU[35] and even the level of EU member states.[36] This is due to inherent fundamental differences about the balancing and justice in relation to management of resources, in particular the use of tangible resources which require planning and have the potential to be exhausted.

Regarding humans and resources, Christianity, with the Bible, clearly points to stewardship and (responsibility for) the care of God's creation,[37] to the importance of reason and reasoning,[38] and to individual responsibility to make choices and to

30 Abukabar (2025).

31 Sposato et al. (2017), pp. 1797–1806.

32 Grober (2007).

33 D'Adamo and Lupi (2021), p. 1861.

34 White (2013), pp. 213–217.

35 De Schutter (2008), pp. 203–236.

36 Domański (2004), pp. 377–381.

37 Boaheng and Korankye (2022), pp. 156–165.

38 Stark (2014), pp. 39–40.

bear the consequences for it.[39] Consequently, Christians should be responsible, active and creative[40] and against threats for sustainability, such as the reckless exhaustion of resources[41] and selfishness.[42] Even empirical studies from the current Western culture jurisdictions confirm that positive relationship between the CSR and the adherence to Christianity and Christian values, i.e. pro-Christianity businesses have higher social and environmental responsibility scores.[43]

However different cultures and non-Christian jurisdictions still have the same issue (how to use tangible, and so inevitably exhaustible, resources) and deal with it in their special manner. Consequently, they address the management and distribution of resources in their particular way and can have their own sustainability, even their own legal concept of sustainability. Western civilization, based upon and shaped by Christianity does not have per se an absolute monopoly for the responsibility, planning, trust, respect, solidarity and many other considerations and naturally it is possible to make the search for any roots and regime of sustainability outside of Europe and even out of Western civilization.

In sum, since the exhaustion of tangible resources has acquired a global dimension, it would be remiss to ignore the existence and development of the attitude in Asia or Africa, based on other than Christianity foundation, and strongly influenced by other factors such as colonialism.[44]

Pursuant to interpretations of the Quran, obedience constitutes active elements, while human destiny remains the passive element.[45] It is even argued that Islam with the Quran is less pro reason and free will and more pro obedience oriented than Christianity with the Bible.[46] In addition, nomadism was a prominent feature in pre-Islamic Arabia and even if the Prophet Muhammad and part of his followers came from urban communities, the focus was rather on moving, trading and expanding,[47] than on staying and exploiting an area, a resource, in a more effective, efficient and possible sustainable manner.[48] Further, for centuries, adherents to Islam, it is argued that Muslims, exhibited the focus on obedience[49] and ideology,[50] along with the reluctance to science[51] and pro-active free creativity to manage one's

---

39 Meylahn (2009), p. a131.

40 MacGregor Pelikánová and Beneš (2023), pp. 39–60.

41 Barbiero (2021), pp. 43–78.

42 D'Adamo and Gastaldi (2023), p. 140944.

43 Dimic et al. (2024), pp. 396–413.

44 Kwashirai (2013), pp. 166–196.

45 Mohseni (2024).

46 Stark (2005), pp. 24–29; Stark (2007), pp. 64–66; Stark (2017), pp. 187–207.

47 Von Grunebaum (1966), pp. 64–72.

48 Stark (2009), pp. 12–33.

49 Fitzmaurice (1990).

50 Hefner (1993).

51 Stark (2014), pp. 314–317.

own resources in a democratic and sustainable manner.[52] It is argued that they appeared to prefer authoritarianism over free and creative endeavours,[53] instead of risk taking they rather waited to take over externally developed knowledge, such as Brahmi/Hindu numbers, or technology to be used in agriculture, such as inventions brought by Christian Crusaders to Muslims (saddles, stirrups, horseshoes).[54] Consequently, the charity and brotherhood ideas as well as other very pro-sustainability thoughts included in the Quran have not had a smooth projection into the applied sustainability and CSR practices and the need to assess their Shariah compliance remains until now.[55] Thus, sustainability and CSR initiatives are neither openly induced nor welcomed nor perfectly managed nor properly communicated[56] and so the multi-stakeholder model cannot be fully developed. This hinders the development and application of a mature and advanced legal concept of sustainability.

Most religions and philosophies outside the Judeo-Christianity tradition do not posit a creation at all and there is not a God as the intelligent designer of a rational universe and so they do not exhibit the drive to engage in discovering the secrets of creation.[57] The active and creative approach to manage given resources based on the free will and respect of future appears remote. For example, Confucianism is based on the belief that human beings need to respect the past with tradition, to adhere to a strong moral standard and live a productive and tranquil life of peace. Such a strong moral and ethical dimension is theoretically very pro-sustainable, but its transposition and materialization are hindered by the strong endorsement of hierarchical loyalty, formal knowledge and focus on the past.[58] In sum, the relationship between Confucian culture and sustainability taking advantage of intellectual property assets, in particular innovations, is ambiguous and problematic.[59] At the same time, it is obvious that Confucianism is against the mentioned sustainability threats, and in particular is definitely against the reckless and self-indulging behavior oriented for present indulgence.

Since sustainability is about ensuring the conditions for "acceptable" welfare levels of all people extended across spatial (intra-generational) and temporal (intergenerational) dimensions, it is, per se, very compatible with Buddhism.[60] Namely, Buddhism is about impermanence, suffering and not self (humans have no permanent soul or self) and their intimate connection to existence and about the trio of morality-mediation-wisdom.[61] The concept of re-birth in Buddhism refers to the

---

[52] Lewis (1993).

[53] Moaddel (2002).

[54] Stark (2009), pp. 67–76.

[55] Moosa (2023), pp. 301–311.

[56] Moosa (2023), pp. 303-311

[57] Stark (2014), p. 315.

[58] Yu and Zhao (2023), pp. 59–76.

[59] Koczkás (2023), pp. 411–431.

[60] Daniels (2011), pp. 35–60.

[61] Dhakal (2021), pp. 71–80.

endless cycle, which is unsatisfactory and painful and can be stopped only by the liberation, aka Nirvana, which extinguishes three fires (poisons): greed, hate and ignorance. In sum, Buddhism is definitely against lavish and reckless waste of resources and definitely pro-sustainable consumption. The emergent social and intellectual developments share much in common with the Buddhist view of the nature of reality, especially the Buddhist concern regarding reducing society's metabolism and on the intrinsic interconnectedness of human existence and interaction with the broader social and natural worlds.[62] Hence, despite very different roots, the sustainability in Europe, at least regarding certain aspects, might look for inspiration in this arena.

Much more complex is the appreciation of the attitude to sustainability in many regions in Africa, because it is heavily marked by the massive colonial and imperial impact,[63] as well as the often abrupt and imposed decolonization leading to injustice.[64] The key feature of the legal concept of sustainability based on Christianity is its ongoing, organic and contextual evolution. However, both colonization and decolonization generated drastic changes with often unpredictable and massive consequences and leading to disorder and mistrust, if not directly chaos. This can be demonstrated via the discourse regarding the Brundtland Report[65] as the (alleged) instrument of "green imperialism" or "eco-colonialism" imposed by Western developed countries onto, poorly prepared for it, "Third World countries,"[66] reacting in an ill manner. If the core values linked to the legal concept of sustainability in Europe are no waste and respect, then these values were boldly violated, i.e. the mentioned drastic changes allegedly in the name of sustainability generated and/or were linked to wasting and disrespecting. Such a "sustainability" is not sustainable.

Legal Documents

Code of Hammurabi, 1755 - 1750 B.C., a collection of 282 rules

Lex Duodecim Tabularum – Duodecim Tabulae (Law of the Twelve Tables), 449 BC

Corpus Iuris Civilis (Codex, Digest or Pandects, Institutes/Institutiones+ Novellae Constitutiones), 529–534

An Act to repeal timber-culture laws, and for other purposes, 16 U.S.C. ch. 2 from 3rd March 1891 ("Forest Reserve Act of 1891")

[62] Daniels (2011), pp. 35–60.

[63] Kwashirai (2013), pp. 166–196.

[64] Domínguez and Luoma (2020), p. 65.

[65] UN Annex to document A/42/427 Report of the World Commission on Environment and Development: Our Common Future from 4 August 1987 ("Brundtland Report").

[66] Bergesen (1988), pp. 6–10.

UN Annex to document A/42/427 Report of the World Commission on Environment and Development: Our Common Future from 4 August 1987 ("Brundtland Report")

References

Abukabar J (2025) The bible and sustainability bringing biblical passages and practices into the ecological debate. Routledge, Oxon

Adamo N, Al-Ansari N (2020) In old Babylonia: irrigation and agriculture flourished under the code of Hammurabi (2000-1600 BC). Earth Sci Geotech Eng 10(3):87–111

Balcerzak A, MacGregor Pelikánová R (2020) Projection of SDGs in codes of ethics—case study about lost in translation? Adm Sci 10(4):95. https://doi.org/10.3390/admsci10040095

Barbiero G (2021) Affective ecology as development of biophilia hypothesis. In: L'Astorina A, Bergami C, De Lazzari A, Falchetti E (eds.) Special Issue "Scientists moving between narratives towards an ecological vision". Visions for Sustainability 16, 5575, 43–78. https://doi.org/10.13135/2384

Bassino JP (2021) Sustainable development in South East Asia. The Cambridge Economic History of the Modern World, pp 146–168. https://doi.org/10.1017/9781316671566.008

Bergesen HO (1988) Reformism doomed to failure? A critical look at the strategy promoted by the Brundtland commission. Int Chall 8(2):6–10

Blackson TA (2011) Ancient Greek philosophy: from the Presocratics to the Hellenistic philosophers. John Wiley & Sons, Inc., Wiley-Blackwell

Boaheng I, Korankye J (2022) A Christian view of stewardship: a study of Daniel 6:1–4. E-Journal Relig Theol Stud (ERATS) 8(6):156–165. https://doi.org/10.38159/erats.2022862

Chiera E (1938) They wrote on clay. University of Chicago Press, Chicago

Copleston F (2003) History of philosophy volume 1: Greece and Rome. Continuum. Doubleday, New York

D'Adamo I, Gastaldi M (2023) Monitoring performance of sustainable development goals in the Italian regions. Sustain For 15:14094. https://doi.org/10.3390/su151914094

D'Adamo I, Lupi G (2021) Sustainability and resilience after COVID-19: a circular premium in the fashion industry. Sustainability 13(4):1861. https://doi.org/10.3390/su13041861

Daniels P (2011) Buddhism and sustainable consumption. In: Zsolnai L (ed) Ethical principles and economic transformation - a Buddhist approach. Issues in Business Ethics, vol 33. Springer, Dordrecht, pp 35–60. https://doi.org/10.1007/978-90-481-9310-3_3

De Schutter O (2008) Corporate social responsibility European style. Eur Law J 14(2):203–236. https://doi.org/10.1111/j.1468-0386.2007.00411.x

Dillon JN (2012) The justice of Constantine. Law, communication, and control. The University of Michigan Press, Ann Arbor

Dimic N, Fatmy V, Vähämaa S (2024) Religiosity and corporate social responsibility: a study of firm-level adherence to Christian values in the United States. Corp Soc Responsib Environ Manag 31(1):396–413. https://doi.org/10.1002/csr.2576

Domański B (2004) West and East in 'New Europe': the pitfalls of paternalism and a claimant attitude. Eur Urban Regional Stud 11(4):377–381. https://doi.org/10.1177/0969776404046272

Domínguez L, Luoma C (2020) Decolonising conservation policy: how colonial land and conservation ideologies persist and perpetuate indigenous injustices at the expense of the environment. Land MDPI 9(3):65. https://doi.org/10.3390/land9030065

Fitzmaurice R (1990) Islam and politics. Studies: an Irish. Q Rev 79(314):177–186. http://www.jstor.org/stable/30091949

Grober U (2007) Deep roots – A conceptual history of 'sustainable development' (Nachhaltigkeit). Best.-Nr. P 2007-002. https://bibliothek.wzb.eu/pdf/2007/p07-002.pdf

Hefner RW (1993) Islam, state, and civil society: ICMI and the struggle for the Indonesian middle class. Indonesia 56:1–35. https://doi.org/10.2307/3351197

Horrell DG, Hunt C, Southgate C (2008) Appeals to the bible in ecotheology and environmental ethics: a typology of hermeneutical stances. Stud Christ Ethics 21(2):219–238. https://doi.org/10.1177/0953946808094343

Johnston D (2011) A brief history of justice. Chapter 3 Aristotle's theory of justice. John Wiley Online Library. https://doi.org/10.1002/9781444397550.ch3

Koczkás S (2023) Confucianism: ancient ideology or driving force of the future? A scoping review on the effect of Confucian culture on innovation. Soc Econ 45(4):411–431. https://doi.org/10.1556/204.2023.00021

Kwashirai VC (2013) Environmental change, control and management in Africa. Glob Environ 12: 166–196

Lewis B (1993) Islam and liberal democracy. The Atlantic. February Issue

MacGregor Pelikánová R (2017) Constantine's Christianity for the (dis)integrated EU - Déjà vu of Constantine's Roman governance reflecting of the mistrial of Jesus for EU? Dialogos 4(1): 81–98. https://doi.org/10.18638/dialogo.2017.4.1

MacGregor Pelikánová R (2023) Accounting science's historical development and its role in social science & accounting theories and their application in accounting sciences. In: Knihová L (ed) Applying interdisciplinarity to globalization, management, marketing, and accountancy science. IGI Global, pp 333. https://doi.org/10.4018/978-1-7998-8669-3.ch009

MacGregor Pelikánová R, Beneš M (2023) A turbulent pathway to uniform patent protection in the EU. Roma Tre Law Rev 2:39–60

MacGregor Pelikánová R, Hála M (2021) CSR unconscious consumption by generation Z in the COVID-19 era – Responsible heretics not paying CSR bonus? J Risk Financ Manage 14(8):390. https://doi.org/10.3390/jrfm14080390

MacGregor Pelikánová R, MacGregor RK (2020) The EU puzzling CSR regime and the confused perception by ambassadors of luxury fashion businesses: a case study from Pařížská. Cent Eur Bus Rev 9(3):74–108. https://doi.org/10.18267/j.cebr.240

MacGregor Pelikanova R, Sani M (2023) Luxury, slow and fast fashion—a case study on the (un)sustainable creating of shared values. Equilibrium Quart J Econ Econ Policy 18(3): 813–851. https://doi.org/10.24136/eq.2023.026

MacGregor Pelikánová R, MacGregor RK, Černek M (2021) New trends in codes of ethics: Czech business ethics preferences by the dawn of COVID-19. Oeconomia Copernicana 12(4): 973–1009. https://doi.org/10.24136/oc.2021.032

Meylahn J-A (2009) Responsibility, god and society: the cry of the other in the sacred texts as a challenge towards responsible global citizenship. HTS Teologiese Stud Theol Stud 65(1):a131. https://doi.org/10.4102/hts.v65i1.131

Moaddel M (2002) The study of Islamic culture and politics: an overview and assessment. Annu Rev Sociol 28:359–386. http://www.jstor.org/stable/3069246

Mohseni MA (2024) Mechanisms of obedience in the holy Quran: a study based on cultural and ethnic semantics and Albert Bandura's social cognitive theory. Quranica - Int J Quranic Res 16(1)

Moosa R (2023) Corporate social responsibility from an Islamic perspective: an overview. Int J Res Bus Soc Sci 12(7):303–311. https://doi.org/10.20525/ijrbs.v12i7.2754

Sposato P, Preka R, Cappellaro R, Cutaia F (2017) Sharing economy and circular economy. How technology and collaborative consumption innovations boost closing the loop strategies. Environ Eng Manag J 16(8):1797–1806. https://doi.org/10.30638/eemj.2017.196

Stark R (2005) The victory of reason. How Christianity led to freedom, capitalism, and Western success. Random House, New York

Stark R (2007) Cities of god. The real story of how Christianity became and urban movement and conquered Rome. Harper One, New York

Stark R (2009) God's battalions. The case for the crusades. Harper One, New York

Stark R (2014) How the west won. ISI Books, Delaware

Stark R (2017) Bearing false witness. SPCK, London

Tkáčová H, Slivka D (2021) The Christian ideal of sustainable development and its perception among young adults: a case study from Slovakia. Religions 12(10):882. https://doi.org/10.3390/rel12100882

US Conference of Catholic Bishops (2012) Compendium: catechism of Catholic Church. 9th printing. USCCB Publishing, Washington. 146:508

Van Tulder R, Van Mil E (2023) Principles of sustainable business. Frameworks for corporate action on the SDGs. Routledge, New York

Von Grunebaum GE (1966) The first expansion of Islam: factors of thrust and containment. Diogenes 14(53):64–72. https://doi.org/10.1177/039219216601405304

White MA (2013) Sustainability: I know it when I see it. Ecol Econ 86:213–217. https://doi.org/10.1016/j.ecolecon.2012.12.020

Yu T, Zhao Z (2023) Confucianism in multicultural China: 'official knowledge' vs marginalised views. Comp Educ 60(1):59–76. https://doi.org/10.1080/03050068.2023.2273641

Chapter 5
The Evolution of Sustainability—From Antiquity to the Twentieth Century

The organization, selection of values, use of intellectual property assets, endorsement of Aristotle's ideas vis-à-vis the future[1] and the multi-spectral commitment to sustainability, with long-term preferences over immediate consumption gratification were among the key factors for the hundreds of years long glory of the Ancient Roman Empire. The disregard of these factors, along with the erosion of the economic, environmental and social commitment contributed to the fall of both Western, and later the Eastern, Roman Empires.[2] In particular, the omnipresent and chronic human self-control issue taking the form of self-indulgence and immediate gratification,[3] in a way that even Romans themselves did not appreciate in the long run, brought an end to both the Western and Eastern Roman Empires.

The following centuries brought not only destabilization and chaos but as well the opportunity for the emergence of a modified version of civilization with newer sustainability concerns in Europe. The Italian Renaissance Republics, the French kingdom, the Czech aristocratic *dominum* and the Hanseatic League added a business and industrial perspective and led to the *Nachhaltigkeit* and *Nachhaltige Entwicklung*, Indeed, sustainability in Europe in the Middle Ages was strongly shaped by the French and Italian influences (Sect. 5.1), while sustainability in New Age Europe was significantly influenced by various central European influences (Sect. 5.2) and ultimately was turned by Germans into *Nachhaltigkeit* (Sect. 5.3), to be challenged by the First Industrial Revolution era (Sect. 5.4), consolidated in the Second Industrial Revolution era (Sect. 5.5) and ultimately made its way in the modern European integration (Sect. 5.6).

[1] Balcerzak and Pelikánová (2020), p. 95.

[2] Harper (2016), pp. 101–111.

[3] O'Donoghue and Rabin (2000), pp. 233–250.

R. MacGregor Pelikánová, *Sustainability in Europe*,
https://doi.org/10.1007/978-3-031-74689-5_5

5.1 Sustainability in Middle Ages Europe: France and Italy

The agriculture-based concept of sustainable development made the transition in Europe in the Middle Ages thanks to the canonic unification, the development of monastic education centers and the rise of universities, in particular those reflecting the Roman heritage incorporated in the *Corpus Iuris Civilis* of the Byzantine Emperor Justinian I. and his famous wife, Theodora.

Charlemagne, called *Pater Europae,* united most of Western Europe for the first time since the classical era of the Roman Empire, and brought, among other innovations, a set of reforms with sustainability impacts. Firstly, his economic and monetary reforms led to the abolishment of the monetary system based on the gold *sou* due to the gold shortage, a move to the silver *denier* and to the standardization via *livre carolinienne* based upon a pound of silver,[4] i.e. one *livre carolinienne* represented one livre of silver and was composed of 20 sous and each sou was composed of 12 deniers and so one *livre carolinienne* was worth 240 *deniers.* Secondly, he instituted principles for the governance of the royal estates and the accounting practice by the *Capitulare de villis* of 802 AD, setting firm rules about the administration of the king's property and how incomes and expenses were to be recorded.[5] Regarding sustainability, it can be pointed out that each of the king's various estates was to be prepared at any given time for the king's use, including having enough food and wine available, but, at the same time, in the case of an excess supply, the surplus that wasn't used due to the king's not visiting the estate was to be sold.[6] Manifestly, a good housekeeping was set in place to avoid waste. Thirdly, he supported the formation of trading communities in the agricultural regions of Provence and the Rhineland. Fourthly, he supported literacy and education. Fifthly, he appreciated the teachings of Augustine of Hippo, who was influenced by Plato, and who recognized the natural law (*lex naturalis*) and questioned the enforceability of unjust man-made law. Sixthly, arguably, already during the era of Charlemagne, the Hindu-Arabic numerals from 1 to 9 made it to Western Europe. Hence, the financial reporting could theoretically be expressed in such numbers already in the ninth century, but it certainly occurred in 976 AD by the *Codex Vigilanus* aka *Codex Albeldensis.* This illuminated compilation of various historical documents from Antiquity to tenth century Spain includes the Hindu-Arabic numerals, which started to spread through Europe thanks to the French cleric, Gerbert of Aurillac, once he became Pope Sylvester II in 999.[7]

Therefore, it may be suggested that the Carolingian Renaissance supported the long-term respect of economic and social aspects, waste avoidance, proper and standardized financial reporting, the enhancement of awareness by education, and the consideration of justice as the ultimate command. More practically, the trading

---

[4]Boline (1953), pp. 5–39.

[5]MacGregor Pelikánová (2023), p. 333.

[6]McKitterick (2008), p. 184.

[7]Freudenhammer (2021), pp. 3–19.

communities and agricultural enterprises of the era were conceived and managed in the cross-generational manner, i.e. as sustainable for several generations locally and even regionally.

As a matter of fact, during the Early Middle Ages, pejoratively labelled by Petrarch as the Dark Ages, a genuine progress in many spheres occurred and it would be misleading and inaccurate to perceive this era simplistically as a time of brutality and backwardness. In contrast, during this era, European civilization had developed massively and sustainability in the areas of agriculture, transport and even production was boosted by a myriad of impressive technological inventions and innovations. Some authorities, including the French historian Jean Gimpel even argue that this was a great innovative era with an unprecedented technology and innovation development.[8] Numerous inventions turned in innovations in mining, metallurgy and similar fields while the manual labor performed originally by slaves was becoming replaced by the use of water and wind power.[9] This all demonstrated the resource effective and efficient, waste avoiding and respect increasing, and thus consequently pro-sustainability, drive by both Romans as well as the "Barbarians" in Europe around the turn of the first millennium.[10]

The reason for the transfer of the knowledge and know-how from the Roman Empire to emerging states in Europe after the upheaval of nations was the maintenance of education and R&D centres despite the chaos. Namely, the practical development launched by Romans could continue thanks to the Catholic invention of Universities, and also monastic intellectual centres. Both of these forms of institutions took advantage of/embodied Western Civilization, which is based on Christianity, stresses the continuing quest for more knowledge so as to better understand the universe, and is intimately linked to individual responsibility, private ownership and creativity.[11] Due to the political development and the crusades, the pro-sustainability agriculture and transport methods and instruments were even exported outside of Europe, see the use of the horse collar, harness and iron horseshoes and adding a wheel to the plow, all of which helped to dramatically increase the effectiveness and efficiency of the agricultural production in territories managed or influenced by the crusaders.[12] In addition, the original Arab agriculture production was done via fragmented small fields, while a much better open-field system had been developed in Northern and Western Europe,[13] i.e. in the homelands of the Crusaders who brought and transposed this method into Arab territories managed or influenced by them.

Going back to France, the Carolingian descendants of Charlemagne were replaced by the Capetian dynasty, of which the direct senior line came to an end in

[8] Stark (2017), pp. 75–77.

[9] Stark (2017), pp. 77–78.

[10] Stark (2014), pp. 90–92.

[11] MacGregor Pelikánová (2017), pp. 81–98.

[12] Stark (2009), pp. 67–70.

[13] Watson (1974), pp. 8–35.

the fourteenth century and the Valois line took over. The last significant member of the direct senior Capetian dynasty was the famed Philip IV the Fair (*Philippe le Bel*) who, from the very beginning of his governance, was determined to consolidate his power and strengthen the monarchy politically, as well as economically, at any cost. He relied on a developed and centralized bureaucracy managed by experts and his consolidation of the Council, Parliament and Court of Auditors were prodigal acts. The Carolingian focus on time extension was complemented by the Capetian focus on space extension, i.e. sustainability was not only the command to maintain and perhaps develop something locally or regionally for several generations, but to do so on the state level. Even more importantly, he forced the sustainability and its regime to cross the legal Rubicon by enacting sustainability related duties, i.e. he could be considered as one of the fathers of the liability for the sustainability. Namely, in 1291, the French king, Philip IV le Bel, was the first monarch who issued a Decree regulating the management of natural resources, while at least indirectly reflecting the sustainable development, see his Edict (*ordonnance*) about the master and mastering of water and forests,[14] i.e. defining *le rôle des maîtres des Eaux et Forêts.*

After the infamous *Tour de Nesle* affair regarding the alleged adultery of his daughters-in-law, Margaret of Burgundy and Blanche of Burgundy, the Capetian direct line became extinct and the governance passed to the Capetian junior line of Valois, namely to Philip VI the Fortunate (*Philippe Le Fortuné*) of Valois. In 1346, this first Valois king issued a Decree called *l'ordonnance de Brunoy* commanding that the maintenance of (the management of) forests be in good shape,[15] i.e. *sur l'administration des forêts, recommandant de les "soutenir en bon état."*

During the eleventh century, *Corpus Iuris Civilis*[16] was discovered in Northern Italy, including the complete text of the *Digest* aka *Pandects,* made up of fifty books,[17] in Pisa or Amalfi.[18] In 1088, history's first university, the University of Bologna,was founded, while its official charter was issued later on by Emperor Frederick Barbarossa. Teachers at this University of Bologna began to lecture based on *Corpus Iuris Civilis* while adding directly their explanatory glosses and comments around the original. Therefore, these early teachers were known as glossators and included the famous four doctors, including Bulgarus, whose pupil was, via Bassianus, the legendary Azo.[19] The importance and influence of summarizing and glossing texts by Azo of Bologna is demonstrated by the famous and even today quoted sentence "*Qui/Chi non ha Azo/Azzo, no vada in Palazzo!*," i.e. no point of attending the trial without referring to writings by Azo of Bologna. Interestingly, Azo of Bologna was not only a law professor and prolific writer, but as well a

[14] *L'ordonnance définissant le rôle des maîtres des Eaux et Forêts,* 1291.

[15] *L'ordonnance de Brunoy*, 1346.

[16] *Corpus Iuris Civilis (Codex, Digest or Pandects, Institutes/Institutiones + Novellae Constitutiones),*529–534.

[17] Osle (2017).

[18] Alburn (1959), pp. 562–642. http://www.jstor.org/stable/25720817.

[19] Pennigton et al. (1995), pp. ix–x.

counsellor of the Commune of Bologna, who was aware about the social and economic development and the potential of lawyers to change the balance of the socio-economic relationship. He attempted to loosen the very strong bonds of feudal society by calling for more liberty of property. i.e. he went with his interpretation of Roman law against feudal customs while distinguishing real and personal rights and, in particular, he was on the side of peasants against landlords.[20] Therefore, the typical agricultural setting for the development of sustainability involved here, not only the economic and environmental pillar, but the social one and this even in the manner with the current perception of the freedom of movement as one of the most fundamental human rights and freedoms.

In the fourteenth century, such a freedom led to an economic boom and to the emergence of the Italian Renaissance in Tuscany, in particular in the cities of Florence and Siena. The economic driving force was the wool industry and the manufacturing and exporting of wool textile products. Hence, the pendulum from the agricultural industry as the leading industry started to move to textile production and trade. As a matter of fact, this can be perceived as a democratic move in the same direction as the French royal management development. Soon, this trend spread from France and Italy to the Netherlands, the German Hansa as well as Central Europe. It was paralleled by dramatic events such as the fall of Constantinople (1453) as well as key inventions such as the mechanical clock (thirteenth century) letterpress printing (1450–1455) and discoveries such as the landing in America (1492). This all made the passage from the Middle Ages to the New Age.

5.2 Sustainability in Europe from the Middle Ages to the New Age: Poland and Bohemia

Around 1450, Johannes Gensfleisch zur Laden zum Gutenberg invented letterpress printing while using his movable-type printing press and issuing copies of poems and in 1455 the famous Gutenberg Bible. This German invention led both to lengthy disputes and a bankruptcy issue for Gutenberg and to the massive spread of literature creating an information revolution. In 1465, the archbishop of Mainz pensioned Gutenberg and stabilized his financial situation by giving him an annual measure of grain, wine, and clothing and exempting him from certain taxes. Interestingly, in 1451 Christopher Columbus was born and he first went to sea about 1465. The key discovery by this Italian explorer, in the service of Spain, happened in 1492, but the promised reward was not met and so Columbus became engaged in a lengthy dispute with the Castilian crown with varying success. Manifestly, being the top inventor or discoverer has never guaranteed a financial success and rather typically leads to drawn-out complicated business and intellectual property law disputes.

[20] Conte (2020), pp. 56–70.

Hence, it is questionable how sustainable were the projects of Gutenberg and Columbus for them, but it is beyond any doubt that they cemented the transition from the Middle Ages to the New Age and led to a new renaissance and to even more modern, and less feudal, business models. Perhaps these were some of the ideas that Azo had in mind 400 years before.

This new manner of business operations and even living conditions reached central Europe and kings, along with bishops and aristocrats, in the renaissance, became engaged in it. In Poland, a perfect example is Nicolaus Copernicus who was a Renaissance man of highly original thinking, who was active as a mathematician, astronomer, physician, statesman, canon of the catholic church, economist and manager.[21] Today, his managerial and inventive activities are often forgotten, as well as the fact that he was a law graduate from the University of Bologna focusing on law, the humanities and astronomy studies. Beginning in 1512, he assumed the administrative and economic duties in Warmia as *magister pistoriae*. During the following decades in Warmia, especially Olsztyn, he made astronomical discoveries, presented a proposal for updating the calendar and published continuously. His most important manuscript for sustainability is a book entitled Locations of Deserted Fiefs (*Locationes mansorum desertorum*) which addresses the issue of populating fiefs with industrious farmers and thus boosting the economy of Warmia in an ongoing manner. Namely, Copernicus came up with pragmatic and sustainable solutions regarding abandoned and poorly managed farms and the lack of investment, e.g. renting to young people, after teaching them the ins and outs of farming, and granting them the bonus of being allowed to make delayed payments.[22] His second most important manuscript for sustainability is a study on the value of money titled On the Minting of Coin or On the Striking of Coin (*Monetae cudendae ratio)*, in which he formulated the monetary rule that "bad" (debased) coinage drives "good" (un-debased) coinage out of circulation. Hence, the Greshman law should be rather called the Copernicus law. In sum, Copernicus's management and recommendations were well recognized in Poland and Prussia and assisted with the stabilization of land management and currency. To put it differently, Copernicus helped to make a sustainable economy and saved monetary policy in both Poland and Prussia. Nevertheless, he was not the only renaissance visionary in central Europe addressing, perhaps unintentionally, all three pillars of sustainability, see e.g. the prodigal terrestrial and hydro-manager Jakub Krčín in Bohemia 50 years later.

Indeed, a very radical and pro-active approach to sustainability was presented by Jakub Krčín z Jelčan a Sedlčan, who was the asset manager of the order of Augustinians and later on of the top Czech aristocratic family, Rožmberks aka Rosenbergs. In 1571, he established the pond Nevděk (now Svět aka World) and in 1584 the pond Rožmberk. He studied at Charles University Prague and probably was exposed to many of the same study materials as Copernicus in Bologna. It is questionable as to whether during his studies, or after, he became exposed to the managerial and

[21] Stachiewicz (1972), pp. 62–81.

[22] Gassendi and Thill (2002), pp. 91–93.

monetary ideas of Copernicus. However, doubtless he continued the work of Štěpánek Netolický in radically changing the South Bohemian landscape by building a set of large ponds to produce, especially, carp fish,[23] i.e. expanding an industry which is still profitable in twenty-first century Bohemia, now the Czech Republic. However, unlike his predecessors, he did not limit his priorities to a mere profitability and productivity, instead he also promoted a systemic rural development.[24] Plainly, it was not only about breeding as many carp fish as possible, but as well about including other considerations, such as the context of other economic (farming, mining), social (employment) and environmental (managing the river flow) factors while performing a rigorous multi-spectral accounting aka *Buchhalting/ Buchhalterie.*

Well, in sum, both Copernicus and Jakub Krčín were definitely influenced by German sources advancing sustainability, as generated by the tradition of Hansa and its trade, and embraced its concepts and terminology. At the same time, they updated this knowledge and so they contributed to the shaping of the famous sustainability German Style—*Nachhaltigkeit.*

5.3 Sustainability German Style: *Nachhaltigkeit*

The Hanseatic League was a rather informal medieval commercial and defensive confederation of merchant guilds and market towns between the 12th and 17th centuries. Ultimately, it included around 170 cities which supported mutual businesses and financed a fleet so as to protect them and their trade against other traders, invaders or pirates. Arguably, it was the first common market in Europe.[25] Undoubtedly, the Hanseatic League was a project based on mutual agreements and arrangements, which was started by German towns and ultimately spread from Bruges to Novgorod, from Stockholm to Krakow, while the centre was Lubeck. Its members, in consideration of their payments, received toll privileges and protection in affiliated communities and their trade routes, including embargo impositions and declarations of war. Although no permanent and independent governance bodies were established, the Hanseatic League developed common trade regulations. This proved to be a weakness and, especially beginning in the sixteenth century, the irregularly negotiating diet was no longer able to effectively and efficiently protect the interests of the merchants from the Hanseatic League towns. In the second half of the seventeenth century, the Hanseatic League disintegrated.

This unique project left behind a set of consequences and impacts which are noticeable even today. Firstly, in the Hanseatic sphere a particular trade uniformity and adaptability occurred leading to the "Europeanization" of Scandinavia and the

[23] Lyach (2023), pp. 197–203.

[24] Čechura (2020).

[25] Haliday (2009).

Baltic.[26] Secondly, the Hanseatic League had accelerated progress and facilitated the move from the Middle Ages to the New Age. Thirdly, Hansa manners of co-operation, trade and management brought many interesting aspects and showed their feasibility. Considering the setting and successful business operations for almost 500 years, which had reflected many democratic rules and principles, the field of sustainability was inevitably impacted and its concept partially transformed and partially organically advanced. The Hansa paved the way to the European common market and ultimately to a single internal market, while respecting the concept of sustainability. As well, this way remained a German theme, i.e. 50 years after the end of Hansa, Germans, inspired by English, French and Italians, came up with the updated concept of sustainability—*Nachhaltigkeit.*

In 1713, the German Colbertist, Hans Carl von Carlowitz followed the Hansa tradition and discussed it in his book *Sylvicultura Oeconomica* in the context of the management of a forest, while focusing on the need to keep producing wood in the same territory and not to produce more wood in more territories aka *nachhaltende Nutzung*.[27] In his endeavours, Carlowitz relied on and further developed in particular two prior models *Sylva or a Discourse of Forest Trees and the Propagation of Timber and His Majesties Dominions* (1662) by John Evelyn and *Grande Ordonnance forestiere sur le fait des Eaux et Forêts*[28] prepared for Luis XIV by Jean Baptist Colbert and promulgated in 1669.[29]

John Evelyn correctly observed that England kept losing woodland and this trend became accelerated during the Civil War. Consequently, he argued, in his best-seller *Sylva*, that such a "furious devastation" must be stopped, the eco-system must be restored, new trees planted, wood imported and fossil fuels used instead of wood. Arguably, "this tenor is very similar to the current UN documents.[30] Around the same time, France, with King Louis XIV faced the same problem—deforestation and the increased need of wood for building the navy and Colbert proposed the *Grande Ordonnance forestiere sur le fait des Eaux et Forêts.* The motto of Colbert "*La France perira faute de bois*" is universal, because the survival of a society without timber and wood is hardly imaginable and it is necessary to re-plant and not waste wood. Basically, Evelyn's strategy and Colbert's Ordonnance targeted the society at large with the aim to induce a general good husbandry regarding forest, wood, timber and their ongoing preservation.

Exactly the same was the starting point for Carlowitz, who was born in an old forest managing family from Saxony. He became the head of mining administration of August I. and his *Sylvicultcura oeconomica* (1713) is a 400 page-long manifest against quick profits causing unrepairable damage. Interestingly, Carlowitz not only brought the term *nachhaltig* in the sense of continuation and sustained usage by

[26] Grohse (2015), pp. 119–123.

[27] von Carlowitz (1713), see the passage about "*Naturmaessige Betrachtung*".

[28] *Grande Ordonnance forestiere sur le fait des Eaux et Forêts*, 1669.

[29] Grober (2007).

[30] Grober (2007).

future generations, but in addition he advanced social ethics. Namely, both Evelyn and Carlowitz kept referring heavily to the Bible, in particular to Genesis, and Carlowitz even moved to consider the lavish, wasteful and harmful use of nature as a sin and to suggest that subsistence must be kept and everybody nourished.[31]

Similarly, in 1832, Emil André addressed in his book *Einfachste den höchsten Ertrag und die Nachhaltigkeit ganz sicher stellende Forstwirthschafts-Methode* the increasing demands ("what was a good war 10 years ago, is not anymore") by underlying the pragmatic approach.[32] The focus was clearly on the continuation and renewal aspect of *Nachhaltigkeit* and *Nachhaltige Entwicklung* within one local territory and one industry. The problematic general feasibility was observed within Malthusian population theory which suggested that the geometric progression of the human population growth, while gaining subsistence with the help of new technologies and other intellectual property assets, can have only the growth in an arithmetical progression, i.e. ultimately natural resources would be exhausted.[33]

In the eighteenth century, the German term *nachhaltig* started to have two meanings—a general meaning and a special meaning referring to a duty in the context of the forest industry, aka a command, to keep resources for future generations—"*die Verpflichtung der Forstwirtschaft, Reserven für künftige Generationen nachzuhalten*."[34] In the nineteenth century, this "*nachhaltige Forstwirtschaft*" was translated as "sustained yield foresty" and this perception led the UN in the twentieth century to the term "sustainable development." This transition and transformation occurred predominantly during the industrial revolution which brought inventions turned into innovations, new business models and an increase in the population. This was a new stage in the development of Sustainability European Style.

5.4 The First Industrial Revolution: The Fight with Little (if any) Sustainability Concerns

When André published his famous book about *Nachhaltigkeit* in 1832 in Prague, the First Industrial Revolution had already started and the transition from production by hand to manufacturing based on automatization and innovations was noticeable. This created a dramatic increase in the effectiveness of production, followed by an unprecedented growth in population and, later on, even an increase in the standard of living. The cradle of the Industrial Revolution was Great Britain, but it spread

[31] Grober (2007).

[32] André (1832) *Einfachste den höchsten Ertrag und die Nachhaltigkeit ganz sicher stellende Forstwirthschafts-Methode*. Borrosh und André, Prague—"*Was vor 10 Jahren gut war, ist es heuter nicht mehr...*".

[33] Dixon and Fallon (1989), pp. 73–78.

[34] Grobe (2013).

promptly to continental Europe and the US and ultimately meant much more than the transition from hand production to manufacturing with machines and the use of water and steam power.

The First Industrial Revolution came with the invention of the steam engine (1765), led to gradual mechanisation and thus was centred around the textile, transportation, coal and steel production. It led to Industry 1.0 (1765–1870) and represented Capitalism 1.0, aka managerial capitalism. It implied a dramatic increase of the output and motion of the growing population which challenged the established social setting. In addition, the first half of the nineteenth century in Europe was a politically extremely unstable era, both on the international as well as emerging national level. The increased manufacturing, boosted by inventions, was launched in the context of massive conflicts, atrocities, wars and chaos, and not confined to only the societal and social levels. In such an unstable setting, the initially agriculture-oriented society gradually transformed into the industrial society, aka Society 1.0, marked by the increasing importance of the secondary sector and growth in human labour productivity[35] along with a migration of the working population into cities.

The standard of living and increase of welfare had not yet been observed, but negative impacts were felt strongly. The production and economy in general during the first half of the nineteenth century was marked by a very low (if any) concern for environmental and social aspects and economic recessions, such as in 1830–1850, led to an often reckless exploitation of resources and taking advantage of not yet fully oriented humans and their labour. On one hand, this was an era of largely unregulated markets with a minimal role for the state and with a firm belief in a simplified version of Adam Smiths's invisible hand and laisser-faire attitude, but, on the other hand, it was as well an era proving Adam Smith' theory of (i) real price of commodities based on labour and (ii) nominal price of commodities based on money. He concluded that that the labourer is rich or poor, is well or ill rewarded, in proportion to the real, not the nominal, price of his labour.[36] His recognition of the importance of labour led him to study the effective and efficient division of labour in order to achieve amicable solutions which could be labelled as sustainable win-win. However, his recognition of the importance of labour, the division of labour and balanced settlement was taken by key economists and philosophers of the nineteenth century and developed in a completely different direction based on their belief about the unsustainable context of the nineteenth century, from economic, social, technological and philosophical perspective.

Indeed, in the same year that An Inquiry into the Nature and Causes of the Wealth of Nations was published (1776), James Watt married notably intelligent Ann MacGregor[37]and thanks to Boulton's financial support,[38] installed one engine for

[35] Turečková et al. (2023), pp. 146–157.

[36] Smith (2012), p. 37.

[37] Jones (1999).

[38] Insley (2011).

pumping water and another for blowing air.[39] Coincidently, in 1776, Johann Wolfgang Goethe became a member of the Weimar Duke's privy council, sat on the highway commission, oversaw the reopening of silver mines, implemented a series of administrative reforms at the University of Jena and engaged in other public activities to find alternative energies, to fight against the plague of forest parasites, and to improve forestry management.[40] So the Scottish contribution of Adam Smith and James Watt to the predominantly economic sustainability got a German touch entailing social and even more environmental aspects.

During this hallmark time, the last and most important representative of German idealism, Georg Wilhelm Friedrich Hegel (1770–1831) was a young boy starting his study at the Gymnasium in Stuttgart. While Adam Smith and James Watt knew, respected and helped each other, it is unclear what was the potential standpoint of Georg Hegel to them. However, it is clear that Georg Hegel met several times Johann Goethe, he became a prolific writer and thinker, and developed ideas about tying together the time and space and justifying the ownership by the existence of the will and independency of humans. Interestingly, James Watt managed to get a patent protection for several of his inventions and, once infringed, patent trials followed (1793–1796) and exactly during this time the first publications of Georg Hegel emerged and one of their results was the creation of one of the intellectual property protection justifications—the meta-physical status of the property created by the author is understood as the embodiment of his will and spirit and so the author or inventor has property rights *in rem*, see the current stream of the perception of moral rights of authors. With a touch of exaggeration, it can be argued that, as much as James Watt fought for the material sustainability of his research and development and inventive activities, to the same extent Georg Hegel fought for the moral sustainability of the personality unity of the will, spirit and their reflections. For sure, 230 years later, we keep struggling regarding the justification of intellectual property, discussing the very meaning of sustainability and spending huge amounts of resources on patent disputes which end with settlements, exactly as happened to James Watt.

Moving back to Georg Hegel, the right wing of his followers, the so called *Althegelianer*, continued with a conservative and pro-Preussen state approach, while the left wing of the so called *Junghegeliane*r departed from the conservatism. Indeed, the massive radicalism of *Junghegelianer* challenged both the religion and societal order, in particular the allegedly corrupt and despotic state apparatus legitimated by Lutheranism, a major branch of Protestantism developing a pietist movement's drive to right social wrongs and focus on individual conversion. This highly self-control and anti-self-indulgence Christianity approach searching for both sustainability and societal peaceful order, including social justice, clashed with more and more industrialized and secularized Europe and with more and more frustrated people exercising their freedom of will and speech. Europe became haunted by a

[39] Scherer (1965).

[40] Rigby (2000), pp. 25–41. http://www.jstor.org/stable/44085757.

spectre—the spectre of communism.[41] Karl Marx and Friedrich Engels took a further step and moved from the initial attack on Christianity to its perception as a mere ideological instrument to obscure society aka "opium of the people" while calling for an organized proletarian revolutionary action and the return of modern society to an allegedly higher form of the most archaic type—collective production and appropriation.[42] Although, Theories of Surplus Value (1861–1863) by Karl Marx has a few points in common or in compliance with Adam Smith's appreciation of the effectiveness and efficiency of labour, as expressed in the An Inquiry into the Nature and Causes of the Wealth of Nations,[43] mostly they are not reconcilable, see e.g. dramatically different perception of the societal evolution and productiveness as well as justice. For example, in this work Karl Marx objected to the view of Adam Smith that the accumulation of capital is achieved by saving, self-denial and immediate gratification denial, but agreed with Hobbes that "Wealth is Power". Ultimately, Karl Marx vulgarized Adam Smith's teachings and argued that "once he had made wages, profit and rent the constituent component parts of exchangeable value or of the total price of the product, he had got himself stuck in the mud and had to get stuck."[44] Plainly, the surplus value (not value added) aka *Mehrwert,* is the new value created by workers in excess of their own labour-cost, which is appropriated by the capitalist as a profit once the product is sold, i.e. a worker by his work adds value to the product, but the worker's wage reflects only a part of this added value transformed into profit, while the rest of such added transformed into profit is kept by the capitalist. Consequently, the gigantic increase in wealth should be due to maximizing surplus value from the employees. Therefore, the importance of sustainability, inventions leading to innovations, other forms of intangible assets, etc., are entirely underplayed by this dogmatic and intellectual property ignoring approach of Marx. As much as we believe in the twenty-first century in technological progress as the most effective way to increase efficiency[45] and even environmental and social effectiveness, so much did Marxists believe in the battle and the ultimate victory of the working class. Nevertheless, the Marxists recognition of the critical importance of human labour, the inherent lack of confidence and aborted justice feeling should be recognized as relevant, even over 150 years after their formulation, see the current crisis of confidence by businesses and the entire capitalist system.[46]

This Marxist turning away from the organic search for co-operation, sharing, improvement, more effectiveness (doing a better job) and more efficiency (doing a job in a better way) by Adam Smith, to a revolutionary abolition of the state,

---

[41] Marx and Engels F (1848) "A spectre is haunting Europe—the spectre of communism. All the powers of old Europe have entered into a holy alliance to exorcise this spectre: Pope and Tsar, Metternich and Guizot, French Radicals and German police-spies."

[42] Marx (1881), p. 346.

[43] Smith (2012), p. 22 et foll.

[44] Marx (1969), p. 103.

[45] Richterová et al. (2021), pp. 553–576.

[46] Royo-Vela and Cuevas Lizama (2022), p. 8505.

including pro-sustainability mechanisms carried by certain capitalists and bourgeoises, needs to be appreciated in the context of the often deplorable social and environmental setting which continued almost until the Second Industrial Revolution.

5.5 The Second Industrial Revolution: The Rediscovery of Sustainability

The Second Industrial Revolution (1870) revolved around a large wave of manufacturing innovations, including inventions regarding the use of electricity (the electric motor in particular), gas and oil and allowed the electrification, assembly line production and a higher division of labour (so much propagated by Adam Smith). Finally, both countries and their populations at large finally experienced economic growth and its positive consequences, including the rise in the living standards making Society 2.0[47] more acceptable to our current standards and the emergence of welfare-social legislation, see the Bismarck's quartet: Insurance Act,[48] Accident Insurance Act,[49] Working Hours Limitation Act and Old Age[50] and Invalidity Insurance Act.[51]

However, this does not mean that Industry 2.0 (1870–1914/1969) brought decades of sustainable peace and prosperity. Indeed, since its beginning, there was observed the negative environment impact, magnification of the effects of monopolies and cartels and the selfish, reckless and extremely anti-sustainable behaviour of conspiring-together oil barons. A legislative action was needed and it was taken on both sides of the Atlantic.

In the United States, in 1876, was created the Federal forest management by the U.S. Department of Agriculture and started a heated two decades long discussion about the exploitative logging and need to protect wooded areas. Consequently, in 1891 Congress passed, and President Benjamin Harrison signed into law, the Forest Reserve Act[52] authorizing the President to designate public lands in the West into "forest reserves." Around the same time, the legislation advanced by the Republican

---

[47] Turečková et al. (2023), pp. 146–157.

[48] *Gesetz, betreffend die Krankenversicherung der Arbeiter (KVG) vom 15.Juni 1883 - Deutsches Reichsgesetzblatt Band 1883, Nr. 9, Seite 73–104* ("German Health Insurance Act 1883").

[49] *Unfallversicherungsgesetz vom 6.Juli 1884 - Deutsches Reichsgesetzblatt Band 1884, Nr. 19, Seite 69* ("German Accident Insurance Act 1884").

[50] *Einschränkung der Frauenarbeit und Kinderarbeit Gesetz von 17.Juni 1887 Deutsches Reichsgesetzblatt Band 1887, Nr. 88, Seite 913* ("Limitation of Working hours Act 1887").

[51] *Gesetz betreffend die Invaliditäts- und Altersversicherung vom 22.Juni 1889 - Deutsches Reichsgesetzblatt Band 1889, Nr. 13, Seite 97* ("German Old Age and Invalidity Insurance Act 1889").

[52] An Act to repeal timber-culture laws, and for other purposes, 16 U.S.C. ch. 2 from 3rd March 1891 ("Forest Reserve Act of 1891").

Senator from Ohio, John Sherman, came to address cartels and their disastrous and totally anti-sustainability impact. He proposed the famous Sherman Antitrust Act, which was enacted in 1890[53] and which meant the official legislative recognition that the simplified version of the invisible hand of the market, from the eighteenth century, needs to be forced. The following decades brought both times of growth in the living standard, improvement of social conditions and even ideas about including ethical concerns in the business conduct[54] as well as the First World War, the Great Depression and the Second World War. This was the time of capitalism 2.0, aka shareholder capitalism, to maximize profits.

The Second Industrial Revolution occurred around the time when *Das Kapital. Kritik der politischen Ökonomie* as a three volume, was published (1867, 1885, and 1894). Marx´ intimated conviction that the inherent contradictions of capitalism will inevitably lead to its decline and collapse became a heavily referred to mantra until the Third Industrial Revolution and partially even beyond. At the same time, it needs to be emphasized that the textile industry became the flagship of the First Industrial Revolution and continued to be important during the Second Industrial Revolution. This was caused by the that fact that the textile industry was one of the first industries going fully for such a (semi)automatic production, the styles dramatically changed and the exponentially growing population, often with small means, needed clothing.[55] As a matter of fact, Friedrich Engels was the son of a wealthy textile manufacturer and the proceeds from this business allowed him to support and subsidize the writings of Karl Marx, such as *Das Kapital.* During the nineteenth century, the upper classes and lower classes started to wear similar outfits and so, instead of rich people wearing expensive clothes and poor people wearing cheap clothes, almost everybody wore similar clothes with the difference in the quantity. The Empire fashion and Victorian aesthetic brought demands for expensive materials, so the resource difference was clearly projected in the selection of materials. Consequently, towards the end of the nineteenth century, clothes looked similar, but rich people had far more of them, and from much more costly materials with expensive accessories. This ultimately led to two fashion streams—(i) *prêt-à-porter* as a wasteful overproduction of standardized clothes and (ii) luxury fashion. Indeed, the textile and fashion industry was both the driving force of the First Industrial Revolution and the Second Industrial Revolution, and the emerging sustainability evil turning out to be the second largest polluter during the Third Industrial Revolution and the Fourth Industrial Revolution.

Indeed, almost all (if not all) industries that were leading lights during the First Industrial Revolution and Second Industrial Revolution are, from the current perspective, perceived as totally unsustainable and in breach of at least two, if not all three, pillars of sustainability. They were about putting machines to work to produce

---

[53] Act to protect trade and commerce against unlawful restraints and monopolies, 15 U.S.C. §§ 1–7, from 2nd July 1890 ("Sherman Antitrust Act 1890").

[54] MacGregor Pelikánová et al. (2021), p. 3715.

[55] MacGregor Pelikánová and Sani (2023), pp. 813–851.

more for more people, regardless of natural resources. Inventions and innovations were about efficiency, but not effectiveness. During these times, it started to become clear that the competitive advantage is about the configuration of resources based on values and market context[56] and these resources are limited. Sustainability concerns were becoming more regional and perhaps even national, but definitely less respectful of nature, as well as certain social classes, than during the prior era of local agriculture.

5.6 From the Third Industrial Revolution towards the Fourth Industrial Revolution and Sustainability in Modern European Integration

The Third Industrial Revolution (1969) revolved around programable logic, nuclear energy and electronics. It led to Industry 3.0 (1969–2000) represented by the automation of production processes and the development of information technology. Society 3.0 was marked by lifestyle changes that have led to the substantial strengthening of economic activities in the tertiary sector.[57] This was paralleled by the emergence of capitalism 3.0, aka customer capitalism with the motto "There is only one valid purpose of a corporation, to create a customer."[58,59] However, already at the very beginning of this era there occurred massive social protests and the rise of left-wing politics along with increased environmental concerns expressed by the Meadows in their "The Limits to growth" (1972) and all this along with the UN endeavours made the pathway leading to the Brundtland Report (1987).

Indeed, after the Second World War, the iron curtain split apart Europe and even the sustainability perceptions. Modern European integration was launched (1951) in order to meet a set of rather heterogenous goals and objectives—to avoid war, to manage more effectively and efficiently resources, to increase competitiveness, to avoid hunger, etc., in Western Europe. After the fall of the Berlin wall (1989), these forces kept spreading even into Eastern Europe, while the world already had entered into the era of strong and intensive globalization. It needs to be emphasized that the time between the Third Industrial Revolution (1969) and the Fourth Industrial Revolution (2000) overlapped, in Europe, not only with the modern integration process and progressive incorporation of the sustainability and sustainable development command in the EU constitutional framework (1997 Amsterdam Treaty) but as well with the eras of the dominance of the shareholder theory (1950–1984), stakeholder theory (1984–1991) and multi-responsibilities theory (1991–2015). In addition, that era saw the pendulum moving from the environmental concerns, see Silent

[56] Moon et al. (2014), pp. 205–225.

[57] Turečková et al. (2023), pp. 146–157.

[58] Drucker (1954).

[59] Drucker (2015).

Spring (1962), The Population Bomb (1968), Limits to Growth (1972), to social concerns, see the Brundtland Report (1987) and, after the Fourth Industrial Revolution, to the famous and alleged balanced and intra-related three pillars structure and SDGs (2015).

Well, the modern era of sustainability started after the Second World War and witnessed a swift movement from Industry 2.0 (1870–1969) over to Industry 3.0 (1969–2000) and then to Industry 4.0 (2000-) with the Internet, renewability, sustainability, and chain processing.[60] Despite the strong leadership of the UN, it can be safely stated that it has strongly shaped Europe, as a matter of fact a deeper review and analysis can show how the development of the Sustainability European Style had accelerated in the last few decades. There were many crises and the EU, at least partially, followed the iconic motto "Never waste a good crisis."[61]

Legal Documents

Corpus Iuris Civilis (Codex, Digest or Pandects, Institutes/Institutiones+ Novellae Constitutiones),529-534

L'ordonnance définissant le rôle des maîtres des Eaux et Forêts, 1291.

L'ordonnance de Brunoy, 1346.

Grande Ordonnance forestiere sur le fait des Eaux et Forêts, 1669

Gesetz, betreffend die Krankenversicherung der Arbeiter (KVG) vom 15.Juni 1883 - Deutsches Reichsgesetzblatt Band 1883, Nr. 9, Seite 73–104 ("German Health Insurance Act 1883")

Unfallversicherungsgesetz vom 6.Juli 1884 - Deutsches Reichsgesetzblatt Band 1884, Nr. 19, Seite 69 ("German Accident Insurance Act 1884")

Einschränkung der Frauenarbeit und Kinderarbeit Gesetz von 17.Juni 1887 Deutsches Reichsgesetzblatt Band 1887, Nr. 88, Seite 913 ("Working Hours Limitation Act 1887")

Gesetz betreffend die Invaliditäts- und Altersversicherung vom 22.Juni 1889 - Deutsches Reichsgesetzblatt Band 1889, Nr. 13, Seite 97 ("German Old Age and Invalidity Insurance Act 1889")

Act to protect trade and commerce against unlawful restraints and monopolies, 15 U.S.C. §§ 1–7, from 2nd July 1890 ("Sherman Antitrust Act 1890")

An Act to repeal timber-culture laws, and for other purposes, 16 U.S.C. ch. 2 from 3rd March 1891 ("Forest Reserve Act of 1891")

[60] Pickard et al. (2019), pp. 30–36.

[61] Van Tulder and Van Mil (2023), pp. xxxi–xxxiii.

References

Alburn CR (1959) Corpus Juris Civilis: a historical romance. Am Bar Assoc J 45(6):562–642. http://www.jstor.org/stable/25720817

André E (1832) Einfachste den höchsten Ertrag und die Nachhaltigkeit ganz sicher stellende Forstwirthschafts-Methode. Borrosh und André, Prague

Balcerzak A, Pelikánová MG (2020) Projection of SDGs in codes of ethics—case study about lost in translation? Adm Sci 10(4):95. https://doi.org/10.3390/admsci10040095

Boline S (1953) Mohammed, Charlemagne and Ruric. Scand Econ Hist Rev 1(1):5–39. https://doi.org/10.1080/03585522.1953.10409898

Čechura J (2020) Jakub Krčín z Jelčan. Vyšehrad, Praha

Conte E (2020) 3. Azo (ca. 1156 to ca.1220/30) and Accursius (1182/5 to ca. 1263). In: Condorelli O, Domingo R (eds) Law and the Christian tradition in Italy. The legacy of the great jurists, 1st edn. Routledge, pp 56–70. https://doi.org/10.4324/9781003014539

Dixon JA, Fallon LA (1989) The concept of sustainability: origins, extensions, and usefulness for policy. Soc Nat Resour 2(1):73–84. https://doi.org/10.1080/08941928909380675

Drucker PF (1954) The practice of management. Harper & Row, New York

Drucker PF (2015) Innovation and Entreprenuership. Routledge Classics, London

Freudenhammer T (2021) Gerbert of. Aurillac and the Transmission of Arabic Numerals to Europe - Gerbert von Aurillac und die Übermittlung der arabischen Ziffern nach Europa Sudhoffs Archiv 105(1):3–19. https://www.jstor.org/stable/48636817

Gassendi P, Thill O (2002) The life of Copernicus (1473–1543): the man who did not change the world. Xulon Press, Fairfax VA, pp 91–93. https://books.google.cz/books?hl=en&lr=&id=9r0RfQtpU6AC&oi=fnd&pg=PA1&dq=academic+articles+copernicus+Locationes+mansorum+desertorum&ots=G6DkkWyIMI&sig=G55VbiPVKnlkEzN7PtJFpPI2Qio&redir_esc=y#v=onepage&q&f=false

Grobe U (2013) Die Entdeckung der Nachhaltigkeit. Kulturgeschichte eines Begriffs. Verlag Antje Kunstmann GmbH, München

Grober U (2007) Deep roots—a conceptual history of 'sustainable development' (Nachhaltigkeit). Best.-Nr. P 2007-002. https://bibliothek.wzb.eu/pdf/2007/p07-002.pdf

Grohse IP (2015) The German Hansa and Bergen 1100-1600 (book review). Scand J Hist 40(1): 119–123. https://doi.org/10.1080/03468755.2014.968754

Haliday S (2009) The first common market? The Hanseatic League History Today 59(7). https://www.historytoday.com/archive/first-common-market-hanseatic-league

Harper K (2016) The environmental fall of the Roman Empire. Daedalus 145(2):101–111

Insley J (2011) Picturing James watt. Br Art J 11(3):37–47. http://www.jstor.org/stable/41615190

Jones PM (1999) Living the enlightenment and the French revolution: James Watt, Matthew Boulton, and their sons. Hist J 42(1):157–182. http://www.jstor.org/stable/3020899

Lyach R (2023) Fisheries and aquaculture in The Czech Republic: history, current status, management, and regulations. Fish Magaz 48(5):197–203. https://doi.org/10.1002/fsh.10913

MacGregor Pelikánová R (2017) Constantine's Christianity for the (dis)integrated EU - Déjà vu of Constantine's Roman governance reflecting of the mistrial of Jesus for EU? Dialogos 4(1): 81–98. https://doi.org/10.18638/dialogo.2017.4.1

MacGregor Pelikánová R (2023) Accounting science's historical development and its role in social science & accounting theories and their application in accounting sciences. In: Knihová (ed) Applying interdisciplinarity to globalization, management, marketing, and accountancy science. IGI Global, p 333. https://doi.org/10.4018/978-1-7998-8669-3.ch009

MacGregor Pelikánová R, Sani M (2023) Luxury, slow and fast fashion—a case study on the (un)sustainable creating of shared values. Equilibrium. Quart J Econ Econ Policy 18(3): 813–851. https://doi.org/10.24136/eq.2023.026

MacGregor Pelikánová R, Němečková T, MacGregor RK (2021) CSR statements in international and Czech luxury fashion industry at the onset and during the COVID-19 pandemic—slowing down the fast fashion business? Sustainability 13(7):3715. https://doi.org/10.3390/su13073715

Marx K (1881) First draft of the letter to Vera Zasulich. Marx-Engels 'Collected Works' 24:346
Marx K (1969) Theories of surplus value. Lawrence & Wishart, London
Marx K, Engels F (1848) Manifesto of Communist Party https://doi.org/10.1504/EJIM.2014.059583
McKitterick R (2008) Charlemagne: the formation of a European identity. Cambridge University Press, New York, p 184
Moon HC et al (2014) Extending Porter's generic strategies: from three to eight. Eur J Int Manag 8(2):205–225. https://doi.org/10.1504/EJIM.2014.059583
O'Donoghue T, Rabin (2000) The economics of immediate gratification. J Behav Decis Mak 13(2): 233–250
Osle D (2017) Justinian and the Corpus Iuris: an overview. Available at SSRN: https://ssrn.com/abstract=2994134 or https://doi.org/10.2139/ssrn.2994134
Pennigton K, Bellomo M, Cochrane LG (1995) Foreword. In: The common legal past of Europe, 1000–1800. Catholic University of America Press, pp ix–x. https://doi.org/10.2307/j.ctt4cg8mj.3
Pickard M, Grecu I, Grecu G (2019) Sustainable smart manufacturing in industry 4.0: real-time resource planning, process monitoring, and production control. Econ Manage Financ Mark 14: 30–36. https://doi.org/10.22381/EMFM14320194
Richterová E, Richter M, Sojková Z (2021) Regional eco-efficiency of the agricultural sector in V4 regions, its dynamics in time and decomposition on the technological and pure technical eco-efficiency change. Equilibrium. Quart J Econ Econ Policy 16(3):553–576. https://doi.org/10.24136/eq.2021.020
Rigby K (2000) Freeing the phenomena: Goethean science and the blindness of Faust. Interdiscipl Stud Liter Environ 7(2):25–41. http://www.jstor.org/stable/44085757
Royo-Vela M, Cuevas Lizama J (2022) Creating shared value: exploration in an entrepreneurial ecosystem. Sustainability 14:8505. https://doi.org/10.3390/su14148505
Scherer FM (1965) Invention and innovation in the watt-Boulton steam-engine venture. Technol Cult 6(2):165–187. https://doi.org/10.2307/3101072
Smith A (2012) An inquiry into the nature and causes of the wealth nations. Wordsworth Classics of World Literature
Stachiewicz WM (1972) Copernicus and his work: a biographical sketch. Polish Rev 17(4):62–81. http://www.jstor.org/stable/25777090
Stark R (2009) God's battalions. The case for the Crusades. Harper One, New York, pp 67–70
Stark R (2014) How the west won. ISI Books, Delaware, pp 90–92
Stark R (2017) Bearing false witness. SPCK, London, pp 75–78
Turečková K, Nevima J, Vaňová A, Vitálišová K (2023) Society 4.0: general economic implications. J Eur Econ 22(2):146–157. https://doi.org/10.35774/jee2023.02
Van Tulder R, Van Mil E (2023) Principles of sustainable business. Frameworks for Corporate Action on the SDGs. Routledge, New York
Von Carlowitz HC (1713) Sylvicultura oeconomica. Braun, Leipzig
Watson AM (1974) The Arab agricultural revolution and its diffusion, 700–1100. J Econ Hist 34(1): 8–35. https://doi.org/10.1017/S0022050700079602

Chapter 6
Modern History of Sustainability

The shift of the interest from regional, and one or more industry focused, continuous long-term production to the global, multi-sectoral, eternal production was orchestrated by the UN[1] and closely followed by the EU. In addition, various private and national streams, including influential academic literature, started to link the international and rather public law sustainability to a national and rather private law social responsibility progressing to CSR and ultimately to CSV. Currently, the UN via SDGs and the EU, via various standardization taxonomic and harmonization reporting endeavours, attempts to tie together both ends, i.e. to connect sustainability to CSR/CSV as a win-win synergy.[2] Namely, it is argued that capitalism in general and businesses in particular are in a crisis of confidence, because on one hand the capitalist system has improved people's living standards, but on the other hand has caused environmental and social problems.[3] The appeasement of negative consequences and other ethical, social, and environmental concerns via a passive and negative CSR, is too little too late, i.e. a systematic and collaborative CSV as the advanced and more active and positive version of the CSR is needed and is feasible to be accomplished.[4]

This is appreciated and endorsed by some, while rejected by others, leading to the conceptual fundamental question whether sustainability is a necessary reality or redundant chimera. In particular, what exactly does sustainability European style mean and/or should mean, and whether this fits into the modern European integration.

The answers to these burning questions can be at least partially extracted from the modern history of sustainability, i.e. the review of the approach and materialization of the concept of sustainability from the end of the Second World War until now.

[1] Schüz (2012), pp. 7–15.

[2] Van Tulder and Van Mil (2023), pp. 379–393.

[3] Szmigin and Rutherford (2013), pp. 171–182.

[4] Royo-Vela and Cuevas Lizama (2022), p. 8505.

R. MacGregor Pelikánová, *Sustainability in Europe*,
https://doi.org/10.1007/978-3-031-74689-5_6

Therefore, it needs to be taken into consideration that the original endeavours of the UN (Sect. 6.1), in the context of the progressively globalized world dominated by large multi-national corporations, called to be socially responsible and so opening the way to the shareholder theory (Sect. 6.2) and later on to the stakeholder theory (Sect. 6.3). The Brundtland Report and Agenda 21 accelerated the recognition of various and, *prima facie,* conflicting three sustainability pillars and dimensions, leading to the admission of various responsibilities and their visualization via the responsibility pyramid of Carroll which, later on, was revised and transformed into CSV and its diagrams by Porter & Kramer (Sect. 6.4). This admission of multi-responsibilities allowed the UN to bring forth the inclusive and various values and priorities sharing sustainability, expressed by the Agenda 2030 and its SDGs and endorsed by the EU Political Guidelines 2019–2024 and the New Political Guidelines with Statement for 2024–2029 (Sect. 6.5). In order to enhance the understanding of the parallel evolution of the concept of sustainability and CSR and the appreciation of the following description of all five states of the modern history of sustainability, a summary is visualized via Table 6.1.

Table 6.1 Overview of modern history of sustainability and CSR history

Sustainability—original UN (economic)	1945–1950			
Sustainability v Shareholder Theory (environ.)	1950–1984	CSR 1.0 (defensive) (stick)	1950–1991	1953 Bowen SRB 1972 Meadows Limits to Growth 1980 Friedman Free to Chose 1984 Freeman Strategic Management 1987 Brundtland Report
Sustainability v Stakeholder Theory (social)	1984–1991			
Sustainability v Multi-responsibilities	1991–2015	CSR 2.0 (charitable, promotional) (stick, carrot)	1991–2006	1991 Carroll Pyramid
		CSR 3.0 (strategic & systemic) (carrot, sermons)	2006–2015	2006 Updated Carroll Pyramid 2006 Porter Kramer CSV 2011 Porter Kramer CSV Updated
Sustainability by Agenda 2030	2015-	CSR 4.0 (collaborative, inclusive, merger) (carrot, sermons, symphonies)	2015-	2023 Van Tulder and Van Mil

Source: Own processing based on Van Tulder and Van Mil (2023, pp. 353 & 543)

6.1 Original Approach to Sustainability by the UN (1945–1950)

In 1919 the League of Nations, by the Treaty of Versailles, was created "to promote international cooperation and to achieve peace and security" after the First World War. Although the following decades showed that the League of Nations failed to achieve world peace, it did contribute to the advancement of the rule of law, to the respect of smaller states and nations and to the recognition of the importance of economic and financial stabilization. Plainly, the League of Nations was the precursor of the UN paving the way towards the respect of others.

In 1945, the UN was formed and the UN Charter[5] was ratified by the majority of signatories, including China, France, the USSR, Great Britain and the United States. Chapter IX, International Economic and Social Cooperation of the UN Charter, provides in Art. 55 that, in order to support peace and friendly relations among nations, the UN "shall promote: higher standards of living, full employment, and conditions of economic and social progress and development;... and universal respect for, and observance of, human rights and fundamental freedoms for all without distinction as to race, sex, language, or religion" (Art. 55 UN Charter). Namely, neither the concept of sustainability nor sustainable development was mentioned, but the development towards a (better) standard of living and human rights was clearly advanced. In 1948, this was further cemented by the UDHR,[6] which states that "Everyone has the right to a standard of living adequate for the health and well-being of himself and of his family…" (Art. 25(1) UDHR). Although the UN Charter and UDHR are international public law instruments which do not impose directly enforceable duties, they are at least highly influential and have set in place patterns and priorities relevant not only to typical international law subjects—states and international organizations, but as well to national law subjects, especially if these subjects are large and multi- or international. Indeed, the UN Charter and UDHR started a progressive move reflected by states and their policies and law, as well as by the literature, and paralleled by a shift in preferred economic ideologies. In sum, economic aspects were mixed with political aspects, i.e. the exploitation of natural resources, the rise of materialism and the growth in per capita income represented the primary concerns[7]—this was the era of the beginning of the cold war and the fight for political and economic power, if not dominance. Sustainability has a clear economic dimension and the economic success was the starting point. At the same time, in the international and progressively global context, it started to become clear that this is not all.

[5]Charter of the United Nations signed on 26 June 1945, in San Francisco, at the conclusion of the United Nations Conference on International Organization ("UN Charter").

[6]UN General Assembly Resolution A/RES/217(III) The Universal Declaration of Human Rights from 10 December 1948 ("UDHR").

[7]Purvis et al. (2019), pp. 681–695.

Although an ad hoc meeting of diplomats in Paris in 1947 to discuss the administration of European Recovery Plan of George Marshall led to the formation of the OEEC in 1948, it was promptly realized that the exclusive focus on economic aspects, such as the distribution of proceeds needed to be expanded and readjusted. After all, via the Marshall Plan, the USA provided necessary resources for promoting not only an economic revival, but as well European integration as such.[8] Logically, there followed the creation of the Council of Europe and the replacement of the OEEC by the OECD in 1961. This further opened the intergovernmental door towards sustainable economic growth in a larger sense and for a larger pool of subjects. Well, the 2nd stage was coming up, see Table 6.1.

6.2 Sustainability v. Shareholder Theory (1950–1984)

Almost in parallel, in the area of the literature and national soft law, there emerged the idea of tying the power to the responsibility vis-à-vis subjects, even if they are not states. Namely, the academic literature, predominantly in the USA, came up with the idea to impose an extended version of the reasonable care duty of businesses, i.e. to make businesses responsible for their impact on society. In 1953, this was famously advanced and explained by the influential book Social Responsibilities of the Businessman by the American economist, Howard R. Bowen, which engaged with the power and impact of the largest US businesses on society as a whole.[9] The exercise of such power should be neither totally discretionary nor without any responsibility for its consequences—such a business has to carry the CSR. The literature kept progressively moving to impose duties related to sustainability and sustainable development, not only upon public but as well private subjects. This academic and rather theoretical trend became real once there occurred the grim realization of the objective shrinking (or at least not growing) resources.

Well, there were more grim events and phenomena. The Iron Curtain split apart the Europe, the Cold War demonstrated massive hostilities even between previously friendly European states and social or environment aspects were underplayed. It was only in the late 1960s and early 1970s, that the rise of the modern environmental movement in the West occurred,[10] and was reflected by a myriad of highly influential publications, such as Silent Spring by Rachel Carson (1962) documenting the environmental damage caused by chemicals and The Population Bomb by Paul R. Ehrlich (1968) reviving the Malthusian concerns regarding the exponential population growth, etc. The economic dimension seemed to be taken care of and the pendulum of attention kept moving to the environmental dimension.[11]

[8] Carolan (2008), pp. 51–65.

[9] Bowen (2013).

[10] Tulloch (2013), pp. 100–114.

[11] Dunlap and Mertig (1991), pp. 209–218.

In 1972, the Meadows team published the hallmark book, The limits to growth, arguing that available resources must be balanced with the increasing world population.[12] This idea was closely linked to concerns expressed during the nineteenth century regarding coal as the most important source of energy and to alarms concerning the exhaustion of coal deposits. It was felt that, should the wasteful consumption of coal continue unchanged, England would lose her dominant industrial position.[13] Similarly, in Germany Rudolf Clausius and others also argued that natural resources, such as coal, ought not to be wasted, because they were non-renewable.[14] Plainly, sustainability and sustainable development became a real, tangible issue, even for very pragmatic businesses, and it started to be clear that each subject of society will have to make the decision about its attitude to sustainability and project it into its involvement, such as accepting or rejecting CSR and no wasting of resources. Such an understanding was further supported by international organizations, and it is important to emphasize the endeavours of the OECD complementing those of the UN, see the OECD Guidelines for multinational enterprises introduced in 1976 and last updated in 2023. Interestingly, it was the OECD Guidelines for multinational enterprises which defined CSR and tied it to sustainability while adding a set of recommended principles.[15]

The years between Bowen and Meadows (1953–1972) were further denoted by a shift in preferred economic ideologies. Firstly, postwar Keynesianism was widely accepted, and the low unemployment and low levels of inflation were celebrated as vehicles to a Golden Age of Capitalism. The strong belief that government intervention can stabilize the economy was projected in fiscal policies and monetary policies and the peak was reached in the 1960s. However, in the late 1960s, a new classical macroeconomics movement emerged with a criticism of the Keynesians' assumptions and with a presentation of its own, sophisticated, mathematical modelling. In the 1970s, the Keynesian economic theory, advancing government intervention and focusing on the interaction between savings and investment, was becoming replaced by the neoliberalist economic theory preferring privatization, deregulation, austerity and a reduction in government spending.[16] The general setting was shaped by the call of Milton Friedman for a reduction of government involvement, reflecting the Chicago school and by the call of Friedrich Hayek for even smaller government actions (*laisser faire* with a mere observance of the rule of law) reflecting the Austrian school.[17]

In 1980, Milton Friedman and his wife Rose Friedman published a best-selling book, Free to Choose: A Personal Statement, which advocates the idea that the free market operation is the best choice because it works for the best interests of all

[12] Meadows et al. (1972).

[13] Du Pisani (2006), pp. 83–96.

[14] Du Pisani (2006), pp. 83–96.

[15] Lauesen (2013), pp. 1771–1777.

[16] Balcerzak and MacGregor Pelikánová (2020), p. 95.

[17] MacGregor Pelikánová and Hála (2021), p. 390.

members of a society and brings about a general prosperity. They rejected interventionist government policies and described them as impairing personal freedoms and economic efficiency. This state minimalist approach led to the famous shareholder theory demanding businesses to do legally all possible to maximize the profits for shareholders, later on slightly moderated towards the motto "what is good for business is good for society."[18] The public sustainability projected in such a private social responsibility leads to CSR, but such a CSR is rather elementary and defensive. To put it differently, the Friedman's shareholder theory is for sustainability and CSR, provided that, in this manner, the shareholder value will be protected as a result, e.g., sustainable expense is acceptable if it avoids a fine imposed by the state. However, the scenery of the eloquent and sophisticated dispute between Keynes and Friedman's supporters and the considerations of what was sufficiently good to increase shareholder value dramatically changed in the 1980s, in particular due to the economic growth slowdown and the increasing population, especially in the Third World. The increasingly globalized world witnessed the pressures of poverty and a growing population, making it extremely difficult for developing countries to behave pro-environmentally. The defensive CSR began to be unsustainable in the interconnected world with its mushrooming unsustainable practices.

Indeed, around 1980, the term sustainability as *normative desideratum* has risen to become one of the top priorities of international politics.[19] It was again the UN which stepped in and pushed further by facilitating the establishment of the WCED in 1983. The prime minister of Norway, Dr. Gor Harlem Brundtland was appointed as its chairperson. The primary reasons for the establishment of the WCED were the failures of successive international reports, unequal world development and the degradation of the environment.[20] Four years later, in 1987, the Brundtland Commission published the famous report "Our common future," developing guiding principles for sustainable development as it is generally understood today. The pendulum kept moving and soon the golden era of Friedman's shareholder theory with shareholder value maximization was (at least partially) over. Similarly, the rather vague international environmental trend has been losing its momentum in the middle of the challenging implementation of perhaps way too many heterogenous initiatives.[21] Consequently, the era of top environmentalism started to be overshadowed by social concerns and the social dimension of sustainability kept growing, the 3rd stage was coming up, see Table 6.1.

---

[18] Porter and Ignatius (n.d.).

[19] Pies and Schultz (2023), p. 101246.

[20] Keeble (1988), pp. 17–25.

[21] Viñuales (2013), pp. 3–13.

6.3 Sustainability v. Stakeholder Theory (1984–1991)

By 1984, the early environmental movements had lost their momentum, as the wave of the radical social movements broke and rolled them back.[22] Later on, both the ecological and social critiques of economic development began to interweave with economic development[23] and so the social dimension became, clearly, a sister dimension to the economic dimension and environmental dimension of sustainability.[24]

Friedman's shareholder theory lifted the veil and proclaimed the business in the form of a company or a corporation, to be a mere vehicle for the business conduct of its shareholders. A slightly expanded version of this theory considered as well (passive) investors, and so the title was updated to the stockholder theory. In either case, the chairperson, typically a CEO, was perceived as a direct employee of the business, and as an indirect employee of the stockholders with the job to maximize their profits. Addressing social or environmental needs by such a business/CEO without aiming for the maximization of profits is, in this perspective, a waste of the stockholder's resources, perhaps even a misappropriation of the entrusted funds and functions. The concept of sustainability is not rejected, but the decision about it is strictly deferred to stockholders, i.e. a business should do what it knows best—earn money, and a stockholder should do what he/she wants, based on their free will. The fundamental idea is that positions are separated and autonomous and interference is not acceptable, because the freedom of will and the protection of private property are fundamental values. However, in 1984, while referring to the same fundamental values, especially the autonomy and free will, R. Edward Freeman rejected the shareholder (stockholder) theory and advanced his stakeholder theory, based on the interconnected relationships between a business and its customers, suppliers, employees, investors, communities, etc.,—its stakeholders.

Freeman's stakeholder theory demands that a business considers not only the interests of its owners and investors, but of the entire society and so it has to work, to a certain extent, not just for its stockholders, but it also should create value for all stakeholders, i.e. direct and indirect. Direct stakeholders are subjects who interact with the business directly, regardless of whether they are internal (employees, stockholders, etc.) or externals (suppliers, customers, etc.). Indirect stakeholders are subjects who interact with the business indirectly (government, media, etc.). All of them are impacted by the business, i.e. the business impacts them and, following Bowen's logic—"who impacts has the responsibility for it." Therefore, the business has a responsibility to act while considering its impact on them. The business does not need to work for all stakeholders (this would not be feasible), but must consider them. Thusly, the stakeholder theory demands a balancing of the various interests of various stakeholders and ultimately modifies the business' ability

[22] Van Der Heijden (1999), pp. 199–221.

[23] O'Riordan (1985), pp. 1431–1446.

[24] Barbier (1987), p. 101.

to generate profits and this can be perceived as trade-offs and limits. Freedom of will and protection of private property occurs in the society with interlinking and intralinking, i.e. businesses do not act alone in a vacuum and individual responsibility is the foundation of our civilization. Interestingly, around the same time as the drafting and presentation of Freeman's stakeholder theory, the Brundtland Commission worked on its report and determined to go much further. The public law sustainability was about to again jump far ahead of the private law CSR in many jurisdictions, but definitely not all, see e.g. Scandinavian countries with the Nordic governance "outside-in" model closely linked to the consensual and welfare state approach,[25] to the respect of minority shareholders and employees and the search for shared values.[26] Well the Scandinavian region was ahead and so it is not surprising that the chair of WCED was from a jurisdiction endorsing the Nordic governance model and relying on a consensual political culture, welfare state and business-government-labour union partnership.[27]

In 1987, after 900 days of cataloguing, the Our Common Future Report by Gro Harlem Brundtland, Norwegian Prime Minister and Chair of the WCED, aka the Brundtland report, was presented to the UN General Assembly and to the UN Conference on Environment Development. It was approved as UN Annex to document A/42/427 Report of the World Commission on Environment and Development Report: Our Common Future and then published by the UN through the Oxford University Press. It was both a genuine milestone and a revolutionary document, expanding the concept of sustainability and sustainable development in its reach (regarding both disciplines and territories) and was soon followed by the EU, its states, even some businesses and a growing part of the literature. It's leitmotif was that poverty is an evil in itself and "a major cause and effect of global environmental problems" (Part I Chapter 1 Point 8) while addressing the entire world population. In addition, the Brundtland Report brought out the heavily cited definition of sustainable development, i.e. "Humanity has the ability to make development sustainable to ensure that it meets the needs of the present without compromising the ability of future generations to meet their own needs. The concept of sustainable development does imply limits - not absolute limits but limitations imposed by the present state of technology and social organization on environmental resources and by the ability of the biosphere to absorb the effects of human activities" (Part I Chapter 3 Point 27). Thus, in 1987, under the auspices of the Brundtland Report, calling for "a new era of economic growth—growth that is forceful and at the same time socially and environmentally sustainable", the debate had come full circle: economic growth was no longer the problem, but it was the solution. The idea of a new win–win scenario emerged by recasting the same old economic growth in "socially and environmentally sustainable" colours.[28] Academia promptly followed it and emphasized the

[25] Mähönen and Johnsen (2019), pp. 218–231.

[26] Carson et al. (2015), pp. 17–19.

[27] Carson et al. (2015), p. 21.

[28] Purvis et al. (2019), pp. 681–95.

contextual meaning of sustainability, the multitude of perspectives (social, economic, ecological), the global intra-related dimension and the common drive to achieve human survival, globally and indefinitely.[29]

Whether Freeman meant, with his stakeholder theory, the entire world is questionable, but there is no doubt that the Brundtland Report did by calling for sustainable development of and for all. Specifically, the Brundtland Report stated that critical global environmental problems were primarily the result of the enormous poverty of the South (states) and the non-sustainable patterns of consumption and production in the North (states). Sustainable development was understood as the capacity to evolve eternally towards reaching an increased efficiency of resource use for the benefit of humanity and other species,[30] i.e. the feature of a massive efficiency (not effectiveness) was added. In sum, sustainable development should provide all people with a basic quality of life while protecting the ecosystem and that sustainable development is about human progress at the global level for a long term.[31]

Consequently, it was proposed that the Brundtland Report approached sustainable development as a conjunction of sustainability perceived as durability, and of development, perceived as extensiveness,[32] and turned into a global command resting on three pillars: environmental (planet), social (people) and economic (profit).

Similarly, moving from an international public macro-perspective, regarding sustainability, to a national private micro-perspective regarding social responsibility, in the aftermath of the Brundtland Report, several theories regarding businesses, management of a business and its competitiveness have been competing with each other. The Friedman traditional, aka conventional, shareholder theory to maximize returns to owners kept going along[33] despite its controversiality, even currently a valid normative theory of business ethics, and its full rejection is desired by many but definitely not by all. However, even Freeman's stakeholder theory to take care of (primary) stakeholders while keeping the business in good health seems to have left its golden era behind, but still is validly existing.[34] Namely, along with these two archetypical theories, there has developed another theory which departs even more from the dogma of generating profits and instead focusing on values, see Carroll's CSR model to do good in society without necessarily aiming at profits.[35] As a matter of fact, the journey to such a theory projected to a model not prioritizing profits started just a few years after the publication of the Brundtland Report. Well, the 4th stage was coming up, see Table 6.1.

---

[29] Brown et al. (1987), pp. 713–719.

[30] Harwood (1990), pp. 3–19.

[31] Sterling (2010), pp. 511–528.

[32] Duran et al. (2015), pp. 806–811.

[33] Friedman (2007), pp. 173–178.

[34] Freeman (2017), pp. 449–465.

[35] Carroll (1999), pp. 268–295.

6.4 Sustainability v. Multi-responsibilities (1991–2015)

The Brundtland report was a clear milestone on the journey towards a multi-spectral sustainability with the universal reach, but still this journey leading to the UN Agenda 2030 was neither shared nor trusted by all.[36] As a matter of fact, the origins of the 'three-pillar' (environmental, economic and social) paradigm of sustainability have been attributed to the Brundtland Report from 1987 as well as to Agenda 21 from 1992, and the 2002 World Summit on Sustainable Development,[37] yet in none of these documents is a clear framework or a theoretical background made explicit.[38] The conceptual struggling in the years following the Brundtland report was marked by the recognized need to balance and make trade-offs between seemingly equally desirable goals within these three categories and this became the prevailing academic tenor.[39] The presence of various values, priorities and consequences from these three categories became obvious and logically led to the identification of various responsibilities and to an ambiguity about how to understand them and become classified.

In 1991, Archie B. Carroll tied together power and consequences and visualized the result by the famous four-layers pyramid, with one responsibility on each layer (economic, legal, ethical and discretionary aka philanthropic). These four-layers responsibilities should overcome any apparent and/or alleged conflicts of a multi-stakeholder model.[40] Firstly, a business has the economic responsibility to be profitable, this is required by society and this is clearly the *conditio sine qua non*—a business without any prospects of ever generating a profit is not sustainable. Secondly, a business has a legal responsibility to obey the law, comply with and fit in the legal framework, and again this is required by society—a business breaching the law on an ongoing basis does not belong in a state with the rule of law. Thirdly, a business has an ethical responsibility to distinguish between good and bad and to do what is just and fair and to avoid harm. Unlike the economic responsibility and legal responsibility, this is not and cannot be required by society. However, this is expected by society and often this expectation can attain the strength of economic and legal responsibilities, see e.g., how certain businesses are tabooed by their customers because they engage in some shady (but not strictly) illegal practices. Fourthly, a business has a philanthropic responsibility which is a mere wish generated by the society in the hope that the business will be a good corporate citizen and as such will show its generosity and concern for others. This pyramid is a truly iconic depiction of social responsibility which is well recognized and updated on an ongoing basis, e.g., by recognizing the overlap of layers and the importance of the ethical layer. The study of the top layers—ethical and philanthropic paved the venue

36 Harlow et al. (2013), pp. 270–280.

37 Moldan et al. (2012), pp. 4–13.

38 Purvis et al. (2019), pp. 681–695.

39 Giddings et al. (2002), pp. 187–196.

40 Friske et al. (2020), pp. 138–155.

to sharing values, aka CSV, which could be perceived as the, so far, highest stage of CSR.

In addition, there needs to be appreciated the European context, namely the Lisbon Strategy,[41] with a new strategic goal, see "5. The Union has today set itself a new strategic goal for the next decade: to become the most competitive and dynamic knowledge-based economy in the world capable of sustainable economic growth with more and better jobs and greater social cohesion." Hence, it can be argued that, ever since 2000, the EU prioritizes sustainable development and is aware that its achievement requires a more coherent and systemic approach, see "38. A fully decentralized approach will be applied in line with the principle of subsidiarity in which the Union, the Member States, the regional and local levels, as well as the social partners and civil society, will be actively involved, using variable forms of partnership. A method of benchmarking best practices on managing change will be devised by the European Commission networking with different providers and users, namely the social partners, companies and NGOs." Namely, the Lisbon Strategy made the multi-stakeholder and pro-values model instrumental.

In 2006, Michael Porter and Mark Kramer came up with the idea of a win-win scenario in the form of connecting the competitive advantage and corporate responsibility leading to the CSV and CSV models and diagrams further advancing the CSR pyramid.[42] They argued that businesses can create shared value in a co-operative manner and by building clusters that improve the productivity and ultimately improve the effectiveness and efficiency.[43] Their followers added that such CSV and clusters are not only linked to the enhancement of innovation, competitiveness and knowledge sharing[44] but as well to the enhancement of environmental and social outcomes.[45] However, further studies confirmed the critical impact of the regulative and normative institutional framework, which might increase the likelihood of selecting for-profit and mixed-value creation while decreasing the likelihood of non-profit value choices.[46] Plainly CSV, even more than CSR, needs the communication and trust ecosystem.[47] It can prosper only within an appropriate legal and cultural setting free of the pervasive injustice, such as of the current (mis)balance of power leading to distorted political and market setting, ecological debt, etc.[48]

Based on the plain diagram, CSV is the intersection of creating social value (addressing social and environmental objectives), typically CSR, and creating business value (being competitive and potentially profit generating). Based on more advanced diagrams, CSV is the intersection of business opportunities, social needs

---

[41] Lisbon European Council on 23 and 24 March 2000 Presidency Conclusions ("Lisbon Strategy").

[42] Porter and Kramer (2006).

[43] Royo-Vela and Cuevas Lizama (2022), p. 8505.

[44] Arikan (2009), pp. 658–676.

[45] Chertow and Miyata (2011), pp. 266–280.

[46] Tautiva et al. (2023), pp. 405–447.

[47] Kramer and Pfitzer (2016), pp. 80–89.

[48] Manzano et al. (2016), pp. 381–393.

and corporate assets and expertise.[49] Conventionally, three mechanisms are proposed for CSV: (i) reconceiving products and markets, (ii) redefining productivity in the value chain and (iii) enabling local cluster development. Similar to CSR, CSV also is not free from criticism.[50] CSV is a cost and to make it a wise investment, it needs to be properly selected,[51] applied and communicated.

Conceptually, the entire CSR-CSV discourse can be perceived as a sub-issue generated by the eternal search for a recipe for a successful business. It is suggested that the best strategy in this respect is to achieve a unique configuration of resources which reflects the mission, goals, market positioning and environment,[52] i.e. balancing resources and values in a multi-stakeholder setting.

As a matter of fact, the need for a proper choice of value and its genuine and non-hierarchical communications[53] across society cannot be over-emphasized.[54] Arguably, a viable (if not the only one) option is to tie both ends together, i.e. SDGs and the social responsibility endeavours of businesses and individuals in an actionable and mutually respecting open-minded manner[55] and cross-sectorial partnership towards CSV,[56] i.e. CSV is not only about creating and sharing value, but as well about its selection and communication.[57] Otherwise pro-sustainable investment is a mere waste[58] and local action unable to address social and environmental injustice, such as ecological debt, in the global setting.[59]

The growing recognition and popularization of Carroll's pyramid and CSV allowed the UN to approach sustainable development as a conjunction of sustainability perceived as durability and of development perceived as extensiveness[60] and to turn it into a global command resting on three pillars: environmental (planet), social (people) and economic (profit), see Agenda 2030[61] and Paris Climate Agreement.[62] The ripening of soil was in progress and so some businesses and states committed to such an approach even before UN agreements were entered into, see

[49] Porter and Kramer (2011), pp. 62–77.

[50] Porter and Kramer (2014), pp. 149–151.

[51] Al-Jundi et al. (2019), pp. 1–2.

[52] Moon et al. (2014), pp. 205–225.

[53] MacGregor Pelikánová et al. (2024a), pp. 214–239.

[54] Ferraro and Beunza (2018), pp. 1187–1207.

[55] MacGregor Pelikánová and MacGregor (2020), pp. 271–282.

[56] Van Tulder and Van Mil (2023), pp. 379–393.

[57] Schaltegger et al. (2016), pp. 3–10.

[58] Drucker (1973).

[59] Manzano et al. (2016), pp. 381–393.

[60] Duran et al. (2015), pp. 806–811.

[61] UN General Assembly Resolution A/RES/71/1 Transforming our world: the 2030 Agenda for Sustainable Development from 25 September 2015 – People, Planet, Prosperity, Peace, and Partnership ("Agenda 2030").

[62] Paris Agreement under the United Nations Framework Convention on Climate Change from 12 December 2015, Chapter XXVII Environment, 7 d., UN Treaty Series, vol. 3156, p. 7 ("Paris Climate Agreement").

the UK legislation fighting against slavery and trafficking even in supply chains.[63] The latest stage was of modern history of sustainability was definitely coming up, see Table 6.1.

6.5 Sustainability by Agenda 2030 and SDGs (2015-...)

The modern concept of sustainability resting on three pillars implied by a set of UN documents, including the Brundtland Report from 1987 and the Agenda 21 from 1992,[64] became very prominent during the first decade of the twenty-first century. Towards the end of this decade, the EU made a critically important decision about its future—the European Commission of José Manuel Barroso launched a strategy for the period 2010–2020 called Europe 2020,[65] which received support as well by the next European Commission of Jean-Claude Juncker as of 2014. The three priorities of Europe 2020 were "the smart, sustainable, inclusive growth" and the five main targets of Europe 2020 included: (i) the increase of the employment rate (to 75%), (ii) the increase of the investment in R&D (to 3% GDP), (iii) the reduction of greenhouse gas emissions (by 20% and 20-20-20), (iv) the reduction in the number of school dropouts (to 10%), and (v) the reduction in the number of Europeans living in poverty (by 25%). Undoubtedly, the EU, especially the European Commission, had worked hard to meet the demands of Europe 2020 and, on the international level logically co-operated with the UN, which steadily kept their role as the top pro-sustainability engine. This cooperation led to the most recent critical dual milestone of the modern concept of sustainability in 2015.

In 2015, two international inter-governmental agreements had a massive and lasting impact on sustainability.[66] Indeed, they made the concept of sustainability more real and actionable, contributed to the move from the CSR to CSV and entered in the sphere of EU policies into the EU law. These two agreements were Agenda 2030 with its 17 SDGs and 169 targets[67] and the Paris Climate Agreement with the target to limit the increase of global temperatures to below 2 °C above pre-industrial levels.[68] The EU helped the UN to prepare and launch both of them and they both got

[63] Modern Slavery Act 2015 c.30 from 26 March 2015 (United Kingdom).

[64] UN Conference on Environment and Development in Rio de Janeiro Declaration Agenda 21 from 3 to 14 June 1992 ("Agenda 21").

[65] COM(2010) 2020 Final Communication from the Commission: EUROPE 2020 A strategy for smart, sustainable and inclusive growth from 3 March 2010 ("Europe 2020").

[66] Millard (2021), pp. 54–63.

[67] UN General Assembly Resolution A/RES/71/1 Transforming our world: the 2030 Agenda for Sustainable Development from 25 September 2015 – People, Planet, Prosperity, Peace, and Partnership ("Agenda 2030").

[68] Paris Agreement under the United Nations Framework Convention on Climate Change from 12 December 2015, Chapter XXVII Environment, 7 d., UN Treaty Series, vol. 3156, p. 7 ("Paris Climate Agreement").

an immediate endorsement by the EU, see e.g. the EU policy called European action for sustainability.[69]

Although the Europe 2020 was coming to its not very laudatory end, due to the failure to meet its five main targets, the EU commitment to sustainability seemed well established. Exactly this was the setting into which the new European Commission of Ursula von der Leyen promoted its Political Guidelines for 2019–2024, aka, "A Union that Strives for More: My Agenda for Europe" which was approved in 2019 with the drive to move "sustainable development" to an unprecedented dimension—via policies and law while closely observed by the literature. Indeed, all six headline ambitions, aka priorities, of Political Guidelines regarding the period 2019–2024 are closely related to sustainable development, in particular the SDGs of UN Agenda 2030 and Paris Climate Agreement:

- A European Green Deal -Striving to be the first climate-neutral continent by reducing net greenhouse gas emissions by at least 55% by 2030, compared to 1990 levels, and to achieve a climate neutrality by?2050;
- An economy that works for people—Ensuring social fairness and prosperity by advancing the EU's unique social market economy which allows economies to grow and to reduce poverty and inequality;
- A Europe fit for the digital age—Empowering people of a new generation by boosting digitalisation from NextGenerationEU and reach 80% of the population with digital skills;
- A stronger Europe in the world—Reinforcing our responsible global leadership by championing multilateralism and a rules-based global order through a more active role and stronger voice for the EU as a global leader, while ensuring the highest standards of climate, environmental and labour protections;
- Protecting our European way of life—Protecting our citizens and our values by advancing equality, tolerance and social fairness, built upon the rule of?law;
- A new push for European democracy—Nurturing, protecting and strengthening our democracy by recognizing that Europe's strength lies in its unity based on shared democratic values, human rights, independence of press, rule of law, and equality.[70]

The Political Guidelines have the key title "A Union that strives for more" and emphasizes the integration of SDGs. This key strategic, and basically all EU policies determining, instrument goes for "a healthy planet, a new digital world, by bringing people together and upgrading social market economy." However, almost prophetically, on 1st December, 2019, when the first European Commission, under the

[69]COM(2016) 739 Final Communication from the Commission to the European Parliament, the Council, the European Economic and Social Committee and the Committee of the Regions: Next steps for a sustainable European future - European action for sustainability. SWD(2016) 390 final from 22 November 2016 ("European Action for Sustainability").

[70]von der Leyen U (2019) Political Guidelines for the next European Commission 2019–2024: A Union that strives for more. My agenda for Europe. Available from: https://www.europarl.europa.eu/resources/library/media/20190716RES57231/20190716RES57231.pdf.

presidency of Ursula von der Leyen, took office, the coronavirus- SARS-CoV-2 causing Covid-19 was reported in Wuhan and soon caused the world pandemic. After two years of struggle and actions often very remote, if not contradictory, to typically sustainable development behaviour, the Covid-19 pandemic seemed at last overcome. The EGD pathway towards the implementations of SDGs and the Paris Agreement became intensified and new milestones were reached, such as the European Climate Law[71] in order to reach a balance of EU-wide greenhouse-gas emissions and their removal regulated in EU law in 2050 and to achieve a net domestic reduction in greenhouse-gas emissions by at least 55% (compared to 1990 levels) by 2030.

However, on 24th February 2022, the Russian Federation invaded Ukraine in an escalation of the war which began in 2014. Consequently, basically all well-established EU policies were challenged, including the EGD and famous New Common Agriculture Policy aka New CAP, which wants to reconcile agriculture production, rural development and agricultural eco-efficiency in the entire EU[72] and which rather unexpectedly goes through the *déjà vu* phenomena linked to the issue of the "butter mountain and milk lake" (1970–2017) created by the government interventionism in the EU. Well, pursuant to the New Political Guidelines[73] with Statement for 2024–2029,[74] the second European Commission of Ursula von der Leyen is open to re-adjust the prior six objectives as expressed in the Political Guidelines for 2019–2024 and modernize the EGD, see the new Clean Industrial Deal, as well the basic philosophy is not dramatically changed, instead it is made more pragmatic and more concerned with prosperity and competitiveness.[75]

Peter Schumpeter placed the entrepreneur at the centre of progress with a potential to be the key transformation agent making the move to the economic and social wealth.[76] A crisis is both a challenge to the pre-existing setting and an

[71] Regulation (EU) 2021/1119 of the European Parliament and of the Council of 30 June 2021 establishing the framework for achieving climate neutrality and amending Regulations (EC) No 401/2009 and (EU) 2018/1999 ('European Climate Law').

[72] Richterová et al. (2021), pp. 553–576.

[73] Von der Leyen U (2024) Political Guidelines for the next European Commission 2024–2029: Europe's Choice, 18 July 2024. Available from: https://commission.europa.eu/document/e6cd4328-673c-4e7a-8683-f63ffb2cf648_en ("New Political Guidelines").

[74] Von der Leyen U (2024) Statement at the European Parliament Plenary by President Ursula von der Leyen, candidate for a second mandate 2024–2029. Our first priority will be prosperity and competitiveness from 18 July 2024. Available from: https://neighbourhood-enlargement.ec.europa.eu/news/statement-european-parliament-plenary-president-ursula-von-der-leyen-candidate-second-mandate-2024-2024-07-18_en ("Statement for 2024–2029").

[75] von der Leyen U (2024) Statement at the European Parliament Plenary by President Ursula von der Leyen, candidate for a second mandate 2024–2029. Our first priority will be prosperity and competitiveness from 18 July 2024. Available from: https://neighbourhood-enlargement.ec.europa.eu/news/statement-european-parliament-plenary-president-ursula-von-der-leyen-candidate-second-mandate-2024-2024-07-18_en.

[76] Horne and Fichter (2022), p. 131163.

opportunity for its change,[77] namely transformative change. Not only Peter Schumpter,[78] but arguably as well Albert Einstein perceived a crisis as a necessary stimulus for human progress, inventiveness and innovations.[79] The stimuli potential and other aspects can be presented via the iconic motto "Never waste a good crisis,"[80] especially since inventions and innovations are hard to be made and their ultimate success rate is arguably under 20%.[81] Indeed, these stimuli should strive to lead to welcome innovative real results based on right values[82] and these results are to be presented and received accordingly[83]—otherwise it would be an increased cost, or even a waste.[84] Well, sometimes such stimuli are more welcome, sometimes less welcome—plainly, each and every innovation depends upon the human capital and both its access to resources and commitment to go forward.[85] In addition, there is a gap between global law and global justice, between rules and subjects active in advanced capitalism,[86] between socio-environmental impacts.

Well, due to crises and stimuli, or in spite of them, the European Commission has neither recalled nor dramatically changed its Political Guidelines with six headline ambitions, in particular the EGD, based on the conviction that the most serious threat for the EU society and competition is climate change,[87] nor has it moved away from "sustainable development means meeting the needs of the present whilst ensuring future generations can meet their own needs."[88] This wording, taken from the Brundtland Report, the explicit reference to three pillars (economic, environmental and social) and the adherence to 17 SDGs from the UN Agenda 2030 are just the starting points. The EU was instrumental in shaping Agenda 2030 and the Sustainability European Style was definitely shaped by the UN endeavours, just as, vice versa, UN endeavours shaped the Sustainability European Style. However, this top international law and rather macro-economic interaction was intimately linked to the bottom-up national law and micro-economic endeavors of other subjects, typically individuals and private companies, who practically and pragmatically advanced their vision of Sustainability European Style, while professional literature kept trying to reconcile all these streams and linked them to theories and doctrines. Hence, the shareholder theory and defensive CSR personalized by Bowen's and Friedman's

[77] Kovoor-Misra (2009), pp. 494–510.

[78] Schumpeter (1934).

[79] D'Adamo and Lupi (2021), p. 1861.

[80] Van Tulder and Van Mil (2023), pp. xxxi–xxxiii.

[81] Moon et al. (2014), pp. 205–225.

[82] Al-Jundi (2019), pp. 1–2.

[83] Drucker (2015).

[84] Drucker (1973).

[85] Polcyn (2021), p. 6846.

[86] Cardesa-Salzmann and Cocciolo (2019), pp. 437–461.

[87] Jones et al. (2023), p. 4.

[88] European Commission (n.d.) Sustainable development. Available from: https://policy.trade.ec.europa.eu/development-and-sustainability/sustainable-development_en. Accessed 3 October 2023.

responsible businessman and later on to the stakeholder theory and charitable and promotional CSR, visualized by the famous four-layers responsibility pyramid of Carroll,[89] were joined by a new and more bottom-up oriented multi-stakeholder model. This model relies on an active and multi-spectral involvement by all and tries to connect the past and the future by sharing values, aka CSV, and depicted as Kramer's and Porter's intersection diagram. This model is entirely dependent upon the commitment of properly informed stakeholders[90] and so it's *bête noire* is the misinformation,[91] in particular its intentional nasty version—disinformation, taking the form of greenwashing, cherry-picking and other practices parasitizing on the Sustainability European Style.[92] However, it would be superficial to perceive the situation simply as a black and white one and to, thusly, overlook various particularities such as spillovers, see the empirical finding that fake news as advertisements can cause increased interest in product domains.[93]

It needs to be emphasized that, in the EU, decisions about the concept of sustainability and especially its application, regardless of who makes them, have a clear central driver—trust,[94] which is linked to two core values—no waste and respect,[95] and should lead to progress via innovations.[96] The recent EU policy and even legislative endeavours, see the CSDDD,[97] make it clear that voluntary action is the preferred method but, if it does lead to improvements, then at least a reporting duty and due diligence are to be imposed on businesses, their holdings and even supply chains. Arguably, SDGs and CSV should represent two sides of the same coin allowing sustainable development based on the same values and heading in the same direction, provided the trust is established.

In sum, the meaning of the ephemeral term sustainability is strongly dependent on the context in which it is applied and on whether its use is based on a social, economic, or ecological perspective, while the common denominator is the interest in human survival on a global scale and recognition that, for this, both a healthy environment and a stable population are needed.[98]

---

[89] Friske et al. (2020), pp. 138–155.

[90] MacGregor Pelikánová and MacGregor (2020), pp. 271–282.

[91] Bennett and Livingston (2018), pp. 122–139.

[92] de Freitas Netto et al. (2020), p. 19.

[93] Rao (2022), pp. 534–554.

[94] Eger et al. (2019), pp. 141–153.

[95] MacGregor Pelikánová et al. (2024b), pp. 1–29.

[96] MacGregor Pelikánová (2019), pp. 13–33.

[97] Directive (EU) 2024/1760 of the European Parliament and of the Council of 13 June 2024 on corporate sustainability due diligence and amending Directive (EU) 2019/1937 and Regulation (EU) 2023/2859 ("CSDDD" or "CS3D").

[98] Brown et al. (1987), pp. 713–719.

Legal Documents

Charter of the United Nations signed on 26 June 1945, in San Francisco, at the conclusion of the United Nations Conference on International Organization ("UN Charter")

UN General Assembly Resolution A/RES/217(III) The Universal Declaration of Human Rights from 10 December 1948 ("UDHR")

UN Conference on Environment and Development in Rio de Janeiro Declaration Agenda 21 from 3 to 14 June 1992 ("Agenda 21")

Lisbon European Council on 23 and 24 March 2000 Presidency Conclusions ("Lisbon Strategy")

COM(2010) 2020 Final Communication from the Commission: EUROPE 2020 A strategy for smart, sustainable and inclusive growth from 3 March 2010 ("Europe 2020")

Modern Slavery Act 2015 c.30 from 26 March 2015 (United Kingdom)

UN General Assembly Resolution A/RES/71/1 Transforming our world: the 2030 Agenda for Sustainable Development from 25 September 2015 – People, Planet, Prosperity, Peace, and Partnership ("Agenda 2030")

Paris Agreement under the United Nations Framework Convention on Climate Change from 12 December 2015, Chapter XXVII Environment, 7 d., UN Treaty Series, vol. 3156, p. 7 ("Paris Climate Agreement")

COM(2016) 739 Final Communication from the Commission to the European Parliament, the Council, the European Economic and Social Committee and the Committee of the Regions: Next steps for a sustainable European future - European action for sustainability. SWD(2016) 390 final from 22 November 2016 ("European Action for Sustainability")

von der Leyen U (2019) Political Guidelines for the next European Commission 2019–2024: A Union that strives for more. My agenda for Europe. Available from: https://www.europarl.europa.eu/resources/library/media/20190716RES57231/20190716RES57231.pdf

Regulation (EU) 2021/1119 of the European Parliament and of the Council of 30 June 2021 establishing the framework for achieving climate neutrality and amending Regulations (EC) No 401/2009 and (EU) 2018/1999 ('European Climate Law')

Directive (EU) 2024/1760 of the European Parliament and of the Council of 13 June 2024 on corporate sustainability due diligence and amending Directive (EU) 2019/1937 and Regulation (EU) 2023/2859 ("CSDDD" or "CS3D")

von der Leyen U (2024) Statement at the European Parliament Plenary by President Ursula von der Leyen, candidate for a second mandate 2024–2029. Our first priority will be prosperity and competitiveness from 18 July 2024. Available from: https://neighbourhood-enlargement.ec.europa.eu/news/statement-european-parliament-plenary-president-ursula-von-der-leyen-candidate-second-mandate-2024-2024-07-18_en ("Statement for 2024–2029")

Von der Leyen U (2024) Political Guidelines for the next European Commission 2024–2029: Europe's Choice, 18 July 2024. Available from: https://commission.europa.eu/document/e6cd4328-673c-4e7a-8683-f63ffb2cf648_en ("New Political Guidelines")

References

Al-Jundi SA, Shuhaiber A, Augustine R (2019) Effect of consumer innovativeness on new product purchase intentions through learning process and perceived value. Cogent Bus Manag 6(1):1–2. https://doi.org/10.1080/23311975.2019.1698849

Arikan AT (2009) Interfirm knowledge exchanges and the knowledge creation capability of clusters. Acad Manag Rev 34:658–676. https://doi.org/10.5465/amr.34.4.zok658

Balcerzak A, MacGregor Pelikánová R (2020) Projection of SDGs in codes of ethics – case study about lost in translation? Adm Sci 10(4):95. https://doi.org/10.3390/admsci10040095

Barbier EB (1987) The concept of sustainable economic development. Environ Conserv 14:101. https://doi.org/10.1017/S0376892900011449

Bennett WL, Livingston S (2018) The disinformation order: disruptive communication and the decline of democratic institutions. Eur J Commun 33(2):122–139. https://doi.org/10.1177/0267323118760317

Bowen HR (2013) Social responsibilities of the businessman. University of Iowa Press

Brown BJ, Hanson ME, Liverman DM, Merideth RW (1987) Global sustainability: toward definition. Environ Manag 11:713–719. https://doi.org/10.1007/BF01867238

Cardesa-Salzmann A, Cocciolo E (2019) Global governance, sustainability and the earth system: critical reflections on the role of global law. Transnatl Environ Law 8(3):437–461. https://doi.org/10.1017/S2047102519000098

Carolan B (2008) The birth of the European Union: US and UK roles in the creation of a unified European Community. Tulsa J Comp Int Law 16(1):51–65

Carroll AB (1999) Corporate social responsibility: evolution of a definitional construct. Bus Soc 38(3):268–295. https://doi.org/10.1177/000765039903800303

Carson SG, Øivind H, Sethi SP (2015) From implicit to explicit CSR in a Scandinavian context: the cases of HÅG and Hydro. J Bus Ethics 127(1):17–31. https://doi.org/10.1007/s10551-013-1791-2

Chertow M, Miyata Y (2011) Assessing collective firm behavior: comparing industrial symbiosis with possible alternatives for individual companies in Oahu, HI. Bus Strategy Environ 20:266–280. https://doi.org/10.1002/bse.694

D'Adamo I, Lupi G (2021) Sustainability and resilience after COVID-19: a circular premium in the fashion industry. Sustainability 13(4):1861. https://doi.org/10.3390/su13041861

de Freitas Netto SV, Sobral MFF, Ribeiro ARB et al (2020) Concepts and forms of greenwashing: a systematic review. Environ Sci Europe 32:19. https://doi.org/10.1186/s12302-020-0300-3

Drucker PF (1973) Management: tasks, responsibilities, practices. Harper & Row, New York

Drucker PF (2015) Innovation and Entrepreneurship. Routledge Classics, London and New York

Du Pisani JA (2006) Sustainable development – historical roots of the concept. Environ Sci 3(2):83–96. https://doi.org/10.1080/15693430600688831

Dunlap RE, Mertig AG (1991) The evolution of the US environmental movement from 1970 to 1990: an overview. Soc Nat Resour 4:209–218. https://doi.org/10.1080/08941929109380755

Duran DC, Gogan LM, Artene AE, Duran V (2015) The components of sustainable development - a possible approach. Procedia Econ Finance 26:806–811. https://doi.org/10.1016/S2212-5671(15)00849-7

Eger C, Miller G, Scarles C (2019) Corporate philanthropy through the lens of ethical subjectivity. J Bus Ethics 156(1):141–153. https://doi.org/10.1007/s10551-017-3551-1

European Commission (n.d.) Sustainable development. Available from: https://policy.trade.ec.europa.eu/development-and-sustainability/sustainable-development_en. Accessed 3 Oct 2023

Ferraro F, Beunza D (2018) Creating common ground: a communicative action model of dialogue in shareholder engagement. Organ Sci 29(6):1187–1207. https://doi.org/10.1287/orsc.2018.1226

Freeman RE (2017) The new story of business: towards a more responsible capitalism. Bus Soc Rev 122(3):449–465. https://doi.org/10.1111/basr.121233

Friedman F (2007) The social responsibility of business is to increase its profits. In: Zimmerli WC, Richter K, Holzinger M (eds) Corporate ethics and corporate governance. Springer, Heidelberg, pp 173–178

Friske W, Nikolov A, Hoang P (2020) CSR reporting practices: an integrative model and analysis. J Market Theory Pract 28(2):138–155. https://doi.org/10.1080/10696679.2019.16903977

Giddings B, Hopwood B, O'Brien G (2002) Environment, economy and society: fitting them together into sustainable development. Sustain Dev 10:187–196. https://doi.org/10.1002/sd.199

Harlow J, Golub A, Allenby B (2013) A review of utopian themes in sustainable development discourse. Sustain Dev 21(4):270–280

Harwood RR (1990) The history of sustainable agriculture In: Edwards CA et al (eds) Sustainable farming systems, pp 3–19

Horne J, Fichter K (2022) Growing for sustainability: enablers for the growth of impact startups – a conceptual framework, taxonomy, and systematic literature review. J Clean Prod 349:131163. https://doi.org/10.1016/j.jclepro.2022.131163

Jones A, Sufrin B, Dunne N (2023) EU competition law. text, cases and materials, 8th edn. Oxford University Press, Oxford

Keeble BR (1988) The Brundtland Report: 'Our Common Future'. Med War 4(1):17–25. https://www.jstor.org/stable/45353161

Kovoor-Misra S (2009) Understanding perceived organizational identity during crisis and change: a threat/opportunity framework. J Organ Change Manag 22:494–510. https://doi.org/10.1108/09534810910983460

Kramer MR, Pfitzer MW (2016) The ecosystem of shared value. Harv Bus Rev 94:80–89

Lauesen LM (2013) OECD Principles of Corporate Governance and CSR. In: Idowu SO, Capaldi N, Zu L, Gupta AD (eds) Encyclopedia of corporate social responsibility. Springer, Heidelberg, pp 1771–1777. https://doi.org/10.1007/978-3-642-28036-8_25

MacGregor Pelikánová R (2019) R&D expenditure and innovation in the EU and selected member states. JEMI – J Entrep Manag Innov 15(1):13–33. https://doi.org/10.7341/20191511

MacGregor Pelikánová R, Hála M (2021) CSR unconscious consumption by Generation Z in the COVID-19 era – responsible heretics not paying CSR bonus? J Risk Financ Manag 14(8):390. https://doi.org/10.3390/jrfm14080390

MacGregor Pelikánová R, MacGregor RK (2020) The willingness of generation Z to officially support CSR – a central European study. DANUBE: Law Econ Soc Issues Rev 11(4):271–282. https://doi.org/10.2478/danb-2020-0016

MacGregor Pelikánová R, Nesbitt T, Balcerzak AP, Oulehla J (2024a) (In)effective communication about social responsibility? Examining large European businesses in the Czech Republic. Bus Manag Econ Eng 22(2):214–239. https://doi.org/10.3846/bmee.2024.21414

MacGregor Pelikánová R, Sani M, Rubáček F (2024b) Sustainable and responsible creation of shared values in the fast fashion industry – I will believe it when I see it. Vis Sustain 21(8873): 1–29. https://doi.org/10.13135/2384-8677/8873

Mähönen J, Johnsen G (2019) Law, culture and sustainability: corporate governance in the Nordic countries. In: Sjåfjell B, Bruner CM (eds) The Cambridge handbook of corporate law, corporate governance and sustainability. Cambridge law handbooks. Cambridge University Press, pp 218–231. https://doi.org/10.1017/9781108658386.022

Manzano JJI, Cardesa-Salzmann A, Pigrau A, Borràs S (2016) Measuring environmental injustice: how ecological debt defines a radical change in the international legal system. J Polit Ecol 23(1): 381–393. https://doi.org/10.2458/v23i1.20225

Meadows DH, Meadows DL, Randers J, Behrens WW (1972) The limits to growth. Universe Books, New York

Millard E (2021) How Fair Trade can stay strong as companies scale up their social and environmental programmes. J Fair Trade 3(1):54–63. https://doi.org/10.13169/jfairtrade.3.1.0054

Moldan B, Janoušková S, Hák T (2012) How to understand and measure environmental sustainability: indicators and targets. Ecol Indic 17:4–13. https://doi.org/10.1016/j.ecolind.2011.04.033

Moon HC et al (2014) Extending Porter's generic strategies: from three to eight. Eur J Int Manag 8(2):205–225. https://doi.org/10.1504/EJIM.2014.059583

O'Riordan T (1985) Research policy and review 6. Future directions for environmental policy. Environ Plan A: Econ Space 17:1431–1446. https://doi.org/10.1068/a171431

Pies I, Schultz FC (2023) The governance of sustainable business model innovation – an ordonomic approach. Scand J Manag 39(1):101246. https://doi.org/10.1016/j.scaman.2022.101246

Polcyn J (2021) Eco-efficiency and human capital efficiency: example of small- and medium-sized family farms in selected European countries. Sustainability 13:6846. https://doi.org/10.3390/su13126846

Porter ME, Ignatius A (n.d.) Creating shared value: an HBR interview with Michael Porter (1/2). Available from http://www.youtube.com/watch?v=F44G4B2uVh4. Accessed 24 Feb 2024

Porter ME, Kramer MR (2006) Strategy and society: the link between competitive advantage and corporate social responsibility. Harv Bus Rev, December

Porter ME, Kramer MR (2011) CSR creating shared value – how to reinvent capitalism – and unleash a wave of innovation and growth. Harv Bus Rev 89:62–77

Porter ME, Kramer MR (2014) A response to Andrew Crane et al.'s article. Calif Manag Rev 56: 149–151

Purvis B, Mao Y, Robinson D (2019) Three pillars of sustainability: in search of conceptual origins. Sustain Sci 14(3):681–695. https://doi.org/10.1007/s11625-018-0627-5

Rao A (2022) Deceptive claims using fake news advertising: the impact on consumers. J Market Res 59(3):534–554. https://doi-org.library.esc.edu/10.1177/00222437211039804

Richterová E, Richter M, Sojková Z (2021) Regional eco-efficiency of the agricultural sector in V4 regions, its dynamics in time and decomposition on the technological and pure technical eco-efficiency change. Equilibrium. Q J Econ Econ Policy 16(3):553–576. https://doi.org/10.24136/eq.2021.020

Royo-Vela M, Cuevas Lizama J (2022) Creating shared value: exploration in an entrepreneurial ecosystem. Sustainability 14:8505. https://doi.org/10.3390/su14148505

Schaltegger S, Hansen EG, Lüdeke-Freund F (2016) Business models for sustainability: origins, present research, and future avenues. Organ Environ 29(1):3–10. https://doi.org/10.1177/1086026615599806

Schumpeter JA (1934) The theory of economic development. Harvard University Press, Cambridge

Schüz M (2012) Sustainable corporate responsibility – the foundation of successful business in the new millennium. Central Eur Bus Rev 1(2):7–15

Sterling S (2010) Learning for resilience, or the resilient learner? Towards a necessary reconciliation in a paradigm of sustainable education. Environ Educ Res 16:511–528

Szmigin I, Rutherford R (2013) Shared value and the impartial spectator test. J Bus Ethics 114:171–182. https://doi.org/10.1007/s10551-012-1335-1

Tautiva JAD, Salvaj Carrera E, Vásquez-Lavín F, Ponce Oliva RD (2023) Understanding the role of institutions and economic context on entrepreneurial value creation choice. Oeconomia Copernicana 14(2):405–447. https://doi.org/10.24136/oc.2023.011

Tulloch L (2013) On science, ecology and environmentalism. Policy Futures Educ 11:100–114. https://doi.org/10.2304/pfie.2013.11.1.100

Van Der Heijden H-A (1999) Environmental movements, ecological modernisation and political opportunity structures. Environ Polit 8:199–221. https://doi.org/10.1080/09644019908414444
Van Tulder R, Van Mil E (2023) Principles of sustainable business. Frameworks for corporate action on the SDGs. Routledge, London and New York
Viñuales JE (2013) The rise and fall of sustainable development. Rev Eur Comp Int Environ Law. Special Issue: Assessing Progress in International Environmental Law 22(1):3–13. https://doi.org/10.1111/reel.12021

Chapter 7
Sustainability in Motion: Sustainable Development

The very wording of the term "sustainable development" suggests that it is a process that creates growth, progress, (positive) change or additions and that such a process can be maintained at a certain rate or level. Typically, in our anthropological perspective, it is assumed that sustainable development is orchestrated by humans and, consequently, it is studied by many different branches of social sciences, such as economics, sociology, legal studies, philosophy or linguistics.

The current understanding of sustainable development must consider its historical and conceptual roots in the evolving context (Sect. 7.1). For the last four decades, pursuant to the majority tenor, the definition of sustainable development has been set by the UN via the Brundtland Report (Sect. 7.2). Although the EU participates in and endorses UN endeavours, including the Brundtland Report, the currently prevailing definition of sustainable development by internal EU sources is slightly different (Sect. 7.3). This ultimately leads to the discussion about what is the current status quo of the perception of sustainable development in the European setting and what are its prospects (Sect. 7.4).

7.1 Historical and Conceptual Roots of Sustainable Development

The term sustainable development emerged hundreds of years ago and has evolved quite a lot during that time. Nevertheless, it was always an ephemeral term facing objections and criticism.[1] The concept of the sustainable development was openly labelled elusive, contradictory[2] and even a cliché.[3] These propositions are currently

[1] Panta (2019), pp. 181–190.

[2] O'Riordan (1996).

[3] Lélé (1991), pp. 607–621.

R. MacGregor Pelikánová, *Sustainability in Europe*,
https://doi.org/10.1007/978-3-031-74689-5_7

suppressed, but still survive below the surface with the vital potential to unexpectedly pop-out in the complex EU crisis[4] and post-crisis setting, regardless of whether from a business[5] or customer perspective.[6]

"Sustainable development" is not a mere neutrally made-up term with a constant meaning, hence the complexity of "sustainable" and "development" grows exponentially when combined into the term "sustainable development." Literally, the "development" is generally understood as an evolutionary process while the "sustainability" is linked to maintaining, in other words keeping, something over time.[7] The term "sustainability" has always represented an ephemeral and evolving concept, perhaps even controversial as a result of conflicting views.[8] Although interesting and relevant considerations for other phenomena and species have been discussed recently,[9] see affective ecology,[10] the majority opinion assigns the key role to the human being, i.e. humans are the engines and targets of sustainable development. Therefore, sustainable development is a subject of global, regional and local interdisciplinarity and multidisciplinary studies entailing views from various branches of science, entailing conventional branches of social science (economics, sociology, legal studies, philosophy or linguistics) as well as other branches (biology, ecology, etc.)[11]

After some hesitations, the current literature settled for the distinction, i.e. it no longer confuses sustainability and sustainable development, and underlines that sustainability tackles systems in order to achieve long-term goals while sustainable development tackles processes addressing human needs and well-being,[12] and which should help to achieve these long-term goals.[13]

As stated above, sustainable development has been for centuries, if not millennia, an important aspect addressed by literature within Europe. This literature stream observed and commented on the reality, i.e. the European drive for development which definitely meets our current expectations regarding sustainability. Indeed, Western Civilization, is shaped by and has foundations in Christianity,[14] is intimately linked to individual responsibility, private ownership and creativity.[15] The practical development orchestrated by Romans exhibiting impressive intellectual property skills brings fruits even nowadays, see, e.g. the communications

[4]Cvik and MacGregor Pelikánová (2021), pp. 109–129.

[5]MacGregor Pelikánová et al. (2021a), pp. 525–546.

[6]MacGregor Pelikánová and Hála (2021), p. 390.

[7]Basiago (1999).

[8]Blewitt (2017).

[9]Barbiero (2016), pp. 28–31.

[10]Barbiero (2021), pp. 43–78.

[11]Feil and Schreiber (2017), pp. 667–681.

[12]Feil and Schreiber (2017), pp. 667–681.

[13]Hector et al. (2014), pp. 7–28.

[14]Stark (2014), pp. 2–5.

[15]MacGregor Pelikánová (2017), pp. 81–98.

infrastructure in Southern Europe. Indeed, the development orientation with sustainability elements could be observed even in what used to be called the Dark Ages and the employment of the newest technologies for transport (e.g., use of a horse collar, harness and iron horseshoes) and agriculture (adding a wheel to the plow) is a perfect example, which, thanks to the crusades, was exported and brought benefits even outside of Europe.[16]

The French command of Philippe IV de Valois for a perpetual renewal of the resource producing fruits perfectly matches up with the German concept and methodology of Hans Carl von Carlowitz and Emil André, leading to the establishment of *Nachhaltige Entwicklung* or *développement soutenable,* aka sustainable development.

The UN Brundtland Report brought out the justice aspect while further developing the idea of using the interest without touching the foundation. However, the literature in the EU has correctly observed the inclusion of ecological, poverty, justice and various ethical issues[17] and pointed to a myriad of juxtapositions and conflicts, making it challenging, if not impossible,[18] to reach a universal model.[19] It is suggested that, similar to the sustainability with three pillars, the "sustainable development" is based on three concepts: (i) development (socio-economic development in line with ecological constraints), (ii) needs (redistribution of resources to ensure the quality of life for all) and (iii) future generations (the possibility of long-term usage of resources to ensure the necessary quality of life for future generations).[20] These are not necessarily in harmony, e.g., often the fight against poverty leads to pollution, etc.[21] As laconically stated, there are hundreds of definitions of sustainable development[22] and even if we settle for the perception of sustainable development as a three pillars structure (matching the sustainability depiction), then we have the issue of the hierarchy, while some argue that the ecological (environmental) dimension is the most important and others reject this as a narrow view.[23]

7.2 Sustainable Development by the UN

The UN was established in 1945 to maintain international peace and security and to develop friendly relations among nations and international cooperation, however, over time, this international and intergovernmental organization expanded its reach

[16] Stark (2009), pp. 67–70.
[17] Rendtorff (2019), pp. 510–524.
[18] Huttmanova and Valentiny (2019), pp. 289–298.
[19] Grabara (2019), pp. 565–570.
[20] Klarin (2018), pp. 67–94.
[21] Chen (2021), pp. 67137–67149.
[22] Dobson (1996), pp. 401–428.
[23] Holden et al. (2014), pp. 130–139.

in the sphere of sustainable development. Other international and intergovernmental projects and endeavours paralleled it and contributed to it, in particular see the intergovernmental association OECD, which, since its establishment in 1960, has been proposing recommendations for economic development and even sustainable economic growth, corporate governance and even CSR.[24]

Starting in 1983, the WCED, led by Dr Gor Harlem Brundtland, conducted research for the UN about the unequal world development and environmental degradation.[25] In 1987, the outcome was presented to the UN General Assembly and to the UN Conference on Environment Development and was approved as a UN Annex to document A/42/427 Report of the World Commission on Environment and Development Report: Our Common Future aka Brundtland Report and then published by the UN through the Oxford University Press. The Brundtland Report brought forth the heavily cited definition of sustainable development, i.e. "Humanity has the ability to make development sustainable to ensure that it meets the needs of the present without compromising the ability of future generations to meet their own needs. The concept of sustainable development does imply limits - not absolute limits but limitations imposed by the present state of technology and social organization on environmental resources and by the ability of the biosphere to absorb the effects of human activities" (Art. I.3.27). This definition entails four important aspects.

Firstly, the definition of sustainable development by the Brundtland Report mentions "humanity" and clearly means the entire world population. No categories, classes, speed-levels, phases, etc., are mentioned and so it is the three Musketeers 'one for all and all for one' approach.

Secondly, it mentions ability which can be understood as the capacity of the entire world population which should ideally work together towards the same goals. To put it differently, the Brundtland Report stated that critical global environmental problems were primarily the result of the enormous poverty of the Southern countries and the non-sustainable patterns of consumption and production in the Northern countries, but at the same time offered a positive collaborative solution.

Thirdly, sustainable development, in the narrow sense, is about satisfying the needs of current and future generations, i.e. it creates an inter-generational respect and perhaps even solidarity.

Fourthly, sustainable development is understood both ethically and pragmatically in the context of available, but progressively diminishing, tangible resources and of the state of technology. Arguably, this might be understood as the recognition of three fundamental values of Western civilization—respect, avoiding waste and creativity.[26]

In sum, the Brundtland Report presents sustainable development as the capacity to evolve eternally while reaching an increased efficiency of resource use for the

---

[24] Lauesen (2013), pp. 1771–1777.

[25] Keeble (1988), pp. 17–25.

[26] MacGregor Pelikánová et al. (2021b), pp. 973–1009.

benefit of humanity and perhaps even other species.[27] Sustainable development is about human progress at a global level for the long term,[28] which means that each and every person should be provided with a basic quality of life while protecting the ecosystem. Such a definition brings with it a myriad of questions and issues leading to controversies and criticism.[29] The Brundtland Report turned sustainable development into a global universal command to go for durable extensiveness,[30] which rests on three pillars: environmental (planet), social (people) and economic (profit), with the concern about the legitimacy of such a command and about the "guaranteed" quality for all.

As already mentioned, various branches of social sciences could be involved in addressing these issues and questions. Economics might discuss the burning issue of distribution and consumption, while sociology might shed light on the social relationships and patterns linked to such a distribution and consumption. Legal studies might deal with the international law foundation and extremely low enforceability on a national level and perhaps, along with philosophy, and, in particular, reality, studying ontology. Last, but not least, linguistics should not be overlooked, especially since some semantically argue that the terms "sustainable" and "development" are incompatible, i.e. one means keeping the status quo and another means moving away from the status quo.

On a more practical level, it has been observed that pre-Brundtland and Brundtland visions and the general drive for pro-environment sustainable development, with its abundance of often uncoordinated, vague, and heterogenous initiatives, face serious implementation issues.[31]

However, perhaps the sharpest conceptual criticism of sustainable development as advanced by the UN, and in particular the Brundtland Report, came from another Norwegian. Gro Harlem Brundtland was the Norwegian leader of the Labour Party and the key person behind the Brundtland Report and its definition of sustainability, while Helge Ole Bergesen was a Norwegian political scientist and politician from the Conservative Party who promptly brought about a massive challenge. Namely, only one year after its issuance, i.e. in 1988 Bergesen started the wave of criticism of the Brundtland Report under the label "green imperialism", aka "eco-imperialism", aka "eco-colonialism", aka "environmental imperialism." The leitmotif of this stream was that the Brundtland Report and its definition of sustainable development are Western developed countries' ideas and concepts imposed on the "Third World."[32] The paternalistic attitude, lack of local appreciation, unrealistic expectations and other features were among those issues that were mentioned.

---

[27] Harwood (1990), pp. 3–19.

[28] Sterling (2010), pp. 511–528.

[29] Harlow et al. (2013), pp. 270–280.

[30] Duran et al. (2015), pp. 806–811.

[31] Viñuales (2013), pp. 3–13.

[32] Bergesen (1988), pp. 6–10.

In 1992, the general perception of sustainable development and the iconic three pillars structure of sustainability, i.e. the economic, environmental and social overlapping dimensions, was confirmed by a Declaration on the UN Conference on Environment and Development in Rio de Janeiro in 1992, aka Agenda 21.[33] Interestingly, Agenda 21 includes a set of principles, such as "Human beings are at the centre of concerns for sustainable development. They are entitled to a healthy and productive life in harmony with nature" (Principle 1) and "All states and all people shall cooperate in the essential task of eradicating poverty as an indispensable requirement for sustainable development, in order to decrease the disparities in standards of living and better meet the needs of the majority of the people in the world" (Principle 5).

The most recent reconfirmation of the UN "Brundtland" version of the definition of sustainable development occurred in 2015 with Agenda 2030.[34] However, already a cursory overview of the Preamble of Agenda 2030 reveals a certain shift, see "This Agenda is a plan of action for people, planet and prosperity. It also seeks to strengthen universal peace in larger freedom. We recognise that eradicating poverty in all its forms and dimensions, including extreme poverty, is the greatest global challenge and an indispensable requirement for sustainable development. All countries and all stakeholders, acting in collaborative partnership, will implement this plan." If the Brundtland Report could be (at least partially and/or theoretically) blamed for an insensibly imposed Western (European) environmentalism, then Agenda 21 and, even more, Agenda 2030, added an increased emphasis on the elimination of poverty and decreasing the disparities in standards of living. The (alleged) prioritization of environmental over social concerns is gone and the most recent perception of sustainable development of the UN reflects the three pillars of sustainability (economic, environmental and social) and the synergic harmony of them as well as of all five Ps: People, planet, prosperity, peace and partnership. Sustainable development should be the process of building bridges between generations (perhaps not limited only to humans) while demanding both individual and collective responsibility,[35] considering the entire biosphere[36] and reflecting moral values going together with a love for life (biophilia).[37] In addition, SDGs and their metrics should be an impulse for the large public, in particular businesses, to embrace a pro-active approach and ultimately develop and implement their own social and green solutions[38] going above and beyond Agenda 2030.

---

[33] Purvis et al. (2019), pp. 681–695.

[34] UN General Assembly Resolution A/RES/71/1 Transforming our world: the 2030 Agenda for Sustainable Development from 25 September 2015 – People, Planet, Prosperity, Peace, and Partnership ("Agenda 2030").

[35] MacGregor Pelikánová et al. (2024), pp. 1–29.

[36] Fitzpatrick (2023), pp. 9–57.

[37] Barbiero (2016), pp. 28–31.

[38] D'Adamo et al. (2024), p. 5049.

What, then, is sustainable development for Europe, and particularly the EU? Is it more pro-environment or more anti-poverty or convinced about the total intra-relation and entire win-win?

7.3 Sustainable Development by the EU

In 1992, the Treaty on European Union[39] was signed by 12 of the member states in Maastricht and it created the EU, while encompassing three separate strands (so-called pillars): the European Communities, a common foreign and security policy, and cooperation between EU governments on justice and home affairs. The original version mentioned progress and a harmonious and balanced development instead of sustainable development, see "The Union shall set itself the following objectives: - to promote economic and social progress which is balanced and sustainable, in particular through the creation of an area without internal frontiers, through the strengthening of economic and social cohesion and through the establishment of economic and monetary union, ultimately including a single currency in accordance with the provisions of this Treaty..." (Art.B Treaty of Maastricht). Accordingly, the Treaty of Maastricht changed the prior Treaty of Rome on EEC, see "The Treaty establishing the European Economic Community shall be amended in accordance with the provisions of this Article, in order to establish a European Community.... Article 2 shall be replaced by the following: Article 2 The Community shall have as its task, by establishing a common market and an economic and monetary union and by implementing the common policies or activities referred to in Articles 3 and 3a, to promote throughout the Community a harmonious and balanced development of economic activities, sustainable and non-inflationary growth respecting the environment, a high degree of convergence of economic performance, a high level of employment and of social protection, the raising of the standard of living and quality of life, and economic and social cohesion and solidarity among Member States." (Art. G Treaty of Maastricht).

In 1997, the Treaty of Amsterdam[40] modified this attitude of the primary EU law by explicitly moving from the original objective "to promote economic and social progress which is balanced and sustainable", as stated by the Treaty of Maastricht in 1992, to the objective "to promote economic and social progress and a high level of employment and to achieve balanced and sustainable development ...". Namely, the EU objective was stated as "Article B The Union shall set itself the following objectives: - to promote economic and social progress and a high level of

[39]Treaty on European Union, signed at Maastricht on 7 February 1992 (92/C 191/01) ("Treaty of Maastricht").

[40]Treaty of Amsterdam amending the Treaty on European Union, the Treaties establishing the European Communities and certain related acts, signed at Amsterdam on 2 October 1997 (97/C 340/01-05) ("Treaty of Amsterdam").

employment and to achieve balanced and sustainable development, in particular through the creation of an area without internal frontiers, through the strengthening of economic and social cohesion and through the establishment of economic and monetary union, ultimately including a single currency in accordance with the provisions of this Treaty" (Art. B of Treaty of Maastricht as revised by the Treaty of Amsterdam).

In 2000, the Lisbon Strategy[41] set new strategic goals for 2010 "5. The Union has today set itself a new strategic goal for the next decade: to become the most competitive and dynamic knowledge-based economy in the world capable of sustainable economic growth with more and better jobs and greater social cohesion." Via the Lisbon Strategy, the European Council involved, in the sustainable development strategy, as well businesses by appealing to their CSR, see "38. A fully decentralised approach will be applied in line with the principle of subsidiarity in which the Union, the Member States, the regional and local levels, as well as the social partners and civil society, will be actively involved, using variable forms of partnership. A method of benchmarking best practices on managing change will be devised by the European Commission networking with different providers and users, namely the social partners, companies and NGOs." This move from substance to process occurred in the context when the EU reinforced its drive for sustainable development while relying on the CSR. This move was motivated by the fear within the EU, that the EU could definitely become outstripped, in global competition, by the USA.[42]

In 2007, the Treaty of Lisbon[43] brought further content and format changes and led to the current version of the TEU, pursuant to which "3. The Union shall establish an internal market. It shall work for the sustainable development of Europe based on balanced economic growth and price stability, a highly competitive social market economy, aiming at full employment and social progress, and a high level of protection and improvement of the quality of the environment. It shall promote scientific and technological advance. It shall combat social exclusion and discrimination, and shall promote social justice and protection, equality between women and men, solidarity between generations and protection of the rights of the child. It shall promote economic, social and territorial cohesion, and solidarity among Member States. It shall respect its rich cultural and linguistic diversity, and shall ensure that Europe's cultural heritage is safeguarded and enhanced" (Art. 3.3 consolidated version of TEU, i.e. ex-Art. 2 TEU).

Regarding the TFEU, its introductory provisions with the general applicability includes the elimination of inequalities (Art. 8 TFEU) and discrimination (Art. 10), social concerns (Art. 9 TFEU), etc. The most important and relevant from them,

[41] Lisbon European Council on 23 and 24 March 2000 Presidency Conclusions ("Lisbon Strategy").

[42] De Schutter (2008), pp. 203–236.

[43] Treaty of Lisbon amending the Treaty on European Union and the Treaty establishing the European Community, signed at Lisbon on 13 December 2007 (2007/C 306/01) ("Treaty of Lisbon").

regarding the sustainable development, is Art. 11 which states "Environmental protection requirements must be integrated into the definition and implementation of the Union's policies and activities, in particular with a view to promoting sustainable development."

Regarding the Charter, the term "sustainable development" is, in a broad sense, mentioned only in the Preamble, while in the operative body it is only mentioned in the context of environmental protection, i.e. "A high level of environmental protection and the improvement of the quality of the environment must be integrated into the policies of the Union and ensured in accordance with the principle of sustainable development" (Art. 37 Charter). Therefore, it might be suggested that sustainable development is rather a homogenous abstract principle on the EU constitutional (primary law) level.

In sum, the EU primary law, in particular the first of this TEU-TFEU-Charter triad, has evolved from (i) promoting balanced and sustainable economic and social progress over to (ii) promoting economic and social progress and sustainable development over to (iii) sustainable development based on balanced economic growth and aiming at social progress and environmental protection. Manifestly, the close co-operation of the EU and the UN, including the work on SDGs, and the EU genuine endorsement of UN concepts and documents, are reflected by the EU primary law. Namely the EU primary law understanding of sustainable development looks, *prima facie*, as compatible with the definition of sustainable development by the Brundtland Report "sustainable development means meeting the needs of the present whilst ensuring future generations can meet their own needs" and Agenda 2030.

Further, on the EU secondary law level, a set of heterogenous and practical, perhaps even casuistic, approaches appear and arguably *prima facie* sustainable development emerges as a very variable tool—regarding accounting harmonization[44] (NFRD,[45] CSRD,[46] etc.), supply chain sustainability (CSDDD aka CS3D),[47]

[44] Balcerzak et al. (2023), pp. 199–212.

[45] Directive 2014/95/EU of the European Parliament and of the Council of 22 October 2014 amending Directive 2013/34/EU as regards the disclosure of non-financial and diversity information by certain large undertakings and groups ("NFRD").

[46] Corporate Social Reporting Directive Directive (EU) 2022/2464 of the European Parliament and of the Council of 14 December 2022 amending Regulation (EU) No 537/2014, Directive 2004/109/EC, Directive 2006/43/EC and Directive 2013/34/EU, as regards corporate sustainability reporting ("CSRD").

[47] Directive (EU) 2024/1760 of the European Parliament and of the Council of 13 June 2024 on corporate sustainability due diligence and amending Directive (EU) 2019/1937 and Regulation (EU) 2023/2859 ("CSDDD" or "CS3D").

sectorial[48] (SFDR)[49] and terminological regulations[50] (Taxonomy Regulation)[51] and a climate neutrality regulation (European Climate Law).[52]

Finally, on the EU supplementary level, a number of influential cases emerged and the CJ EU appeared generally, but not always, inclined to follow the vision of the European Commission while interpreting the EU law vis-à-vis sustainability and sustainable development. In *C-62/88 Greece v. Council (Chernobyl I Case)* and in *C-281/01 Commission v. Council (Energy-efficient labelling),* the CJ EU followed the argumentation of the Commission and took the approach that trade measures, which are in the sphere of exclusive competence, keep their regime even if environmental concerns involved, i.e. trade measures are trade measures whatever the objective pursued and so environmental concerns do not move them into the sphere of shared competence.[53] Therefore, sustainability and sustainable development concerns can by this mechanism move from the shared competence to the exclusive competence regime and the CJ EU does not consider it as the usurpation of power of the EU and/or European Commission. Even if some infringements of the EU law happen, this does not cause, per se, non-contractual liability of the EU, see *T-74/00 Artegodan v Commission (Tenuate Retard)* denying the compensation for the withdrawal of marketing authorization of the medicinal product Tenuate Retard. In sum, there is a large and broad consideration of the EU and EU law to balance the protection of health, environmental protection and consumer protection v the economic interests of traders and, based on the precautionary principle, The EU can take protective measures without having to wait for the reality and the seriousness of those risks to be fully demonstrated, see *C-343/09 Afton Chemical Limited v Secretary of State for Transport* (Greenhouse gas emission monitoring and reduction—labelling "Contains metallic additives"). Even EU member states have a large discretion and can move even further with the EU principle regarding the environment pollution "polluter pays," see *C-378/08 Raffineire Mediterranée Raffinerie Mediterranee (ERG) SpA (Augusta Sicily decontamination).* Indeed, the EU even "exports" to a certain measure its vision of the sustainable development and projects the observation and compliance with the Sustainability European Style in its international trade agreements, see *C-2/15 Opinion of the Court on the Free Trade Agreement between the EU and the Republic of Singapore (criterion of the sustainable development),* in particular sustainable development v renewable energy

[48] Rubáček et al. (2023), pp. 123–143.

[49] Regulation (EU) 2019/2088 of the European Parliament and of the Council of 27 November 2019 on sustainability-related disclosures in the financial services sector ("SFDR").

[50] MacGregor Pelikánová and Rubáček (2022), pp. 173–195.

[51] Regulation (EU) 2020/852 of the European Parliament and of the Council of 18 June 2020 on the establishment of a framework to facilitate sustainable investment, and amending Regulation (EU) 2019/2088 ("Taxonomy Regulation").

[52] Regulation (EU) 2021/1119 of the European Parliament and of the Council of 30 June 2021 establishing the framework for achieving climate neutrality and amending Regulations (EC) No 401/2009 and (EU) 2018/1999 ('European Climate Law').

[53] Leczykiewicz (2005), pp. 1673–1685.

generation. Indeed, the sustainable development is becoming an often considered and discussed phenomenon in the case law of the CJ EU and recently, in particular, the issue of the promotion of the use of energy from renewable sources, see *C-148/23 Gestore dei Servizi Energetici SpA – GSE v Erg Eolica (Renewable Energy Roadmap).*

As a matter of fact, the EU even moved to its (own) definition of sustainable development. However it did not do it via its primary or secondary or supplementary law, but instead via its policies. In 2010, one of the very first steps in this arena was taken by the Communication from Commission COM (2010) 2020 EUROPE 2020 Strategy for smart, sustainable and inclusive growth, aka Europe 2020. In 2016, a genuine determination of sustainable development was presented via Communication COM(2016) 739 "Next steps for a sustainable European future" in the aftermath of Agenda 2030. Its wording is self-explanatory "1. Introduction: the EU's commitment to Sustainable Development. Sustainable development has since long been at the heart of the European project. The EU Treaties give recognition to its economic, social and environmental dimensions which should be addressed together. The EU is committed to development that meets the needs of the present without compromising the ability of future generations to meet their own needs. A life of dignity for all within the planet's limits that reconciles economic prosperity and efficiency, peaceful societies, social inclusion and environmental responsibility is at the essence of sustainable development." As a matter of fact, the Communication COM(2016) 739 "Next steps for a sustainable European future" explicitly and expressly refers to Agenda 2030 and its 17 SDGs, see "2.3. The European Union as a committed global partner to promote the 2030 Agenda. The vision of the 2030 Agenda is fully consistent with the objectives of EU external action, including the pursuit of sustainable development (See Article 21 (2) d) of the Treaty on European Union)." However, as much as the meeting of SDGs appears more and more challenging and often not feasible by 2030, unless the current approaches change dramatically,[54] both outside and inside criticism, regarding Communication COM (2016) 739 Next steps for a sustainable European future, keeps on mushrooming.

Already in 2017 there was presented the Opinion of the European Economic and Social Committee aka EESC on Communication COM(2016) 739 Next steps for a sustainable European future expressing disappointment and blaming the European Commission for an ineffective leadership, insufficient co-operation, lack of a pragmatic drive, and lacking a participatory process and poor coordination, see "3.5.The EESC acknowledges the efforts of the Commission, as reflected in this Communication, to link SDGs with current EU policies and the Commission's ten work priorities. However, the Communication focuses too much on the current business of EU policies. The Commission does not understand the 2030 Agenda as a call to critically review and change its own policies. The Communication does not reflect the paradigm shift that the SDGs stand for: a new model of development that is economically more sustainable, socially more inclusive and environmentally more

[54] MacGregor Pelikánová and Sani (2023), pp. 813–851.

viable over the long term, and that ensures that the resources of our planet will be shared fairly with a growing world population." (3.5 Opinion of EESC).

Well, in 2019, the COM(2019) 22 the Reflection Paper "Towards a Sustainable Europe by 2023" came to address extensively, vigorously and consistently, SDGs and their implementation while repeating that sustainable development is the development that meets the needs of present generations without compromising the ability of future generations to meet their own needs and highlighting that it is deeply rooted in the European project. Further, it provides a rather elaborate analysis of sustainable development. Already on page 2 of the Reflection Paper is included the Foreword by FVP Timmermans and VP Katainen explicitly stating that "Sustainable development is a complex issue, but a simple concept: it is about making sure that our economic growth allows us to maintain a model that produces fair outcomes for all of humanity; and about ensuring that humans don't consume more resources than the Earth has to offer. That means we need to modernise our economy to embrace sustainable consumption and production patterns, to correct the imbalances in our food system, and to put our mobility, the way we produce and use energy, and design our buildings onto a sustainable path. To do this we also need to gear all our science, our financing, taxation, and our governance towards the achievement of the SDGs. This is not a left or right issue, it is a right or wrong issue." The EU pendulum swung from the initial consideration of sustainable development from a macro-perspective to a multi-stakeholder model with a micro-perspective entailing, in particular, businesses, their sustainable supply chains,[55] and more consumers becoming (arguably) more responsible and greener.[56] Interestingly, the legal requirement imposed upon large businesses to disclose their potentially anti-social and anti-environmental behaviour was brought in 2014 by the NFRD[57] into the EU law, but the extension of the scope of this duty to supply chains was brought first in the UK, see the Modern Slavery Act in 2015.[58] However, instead of the synergy of such a pro-sustainable development reporting legislation, the Brexit was launched by the referendum on 23 June 2016 and finalized on 31 January 2020.

Only a few months later, the candidate for the European Commission presidency, Ursula von der Leyen, presented her six top ambitions aka priorities via the Political Guidelines for 2019–2024.[59] The first of these six priorities is the famous COM (2019) 640 final - The European Green Deal, which definitely moved "sustainable development" to a higher level, i.e. the pro-sustainable development era of Jose

[55] Małys (2023), pp. 255–286.

[56] MacGregor Pelikánová (2024), pp. 218–235.

[57] Directive 2014/95/EU of the European Parliament and of the Council of 22 October 2014 amending Directive 2013/34/EU as regards disclosure of non-financial and diversity information by certain large undertakings and groups ("NFRD").

[58] Modern Slavery Act 2015 c.30 from 26 March 2015 (United Kingdom).

[59] Von der Leyen U (2019) Political Guidelines for the next European Commission 2019–2024: A Union that strives for more. My agenda for Europe. Presented to the Members of European Parliament on 16 July 2019. Available from: https://www.europarl.europa.eu/resources/library/media/20190716RES57231/20190716RES57231.pdf ("Political Guidelines").

Manuel Barosso and Jean-Claude Juncker with the Europe 2020 has been replaced, as of 1st December 2019, by an even more progressive pro-sustainable development era of Ursula von der Leyen and her Political Guidelines for 2019–2024 with six priorities, in particular the European Green Deal, as well as her Statement for 2024–2029, in particular the Clean Industrial Deal. Therefore, the exploration of the current meaning of "sustainable development" is tied to these endeavours as well as global justice,[60] energy transition and socio-economic transformation which face the temptation to revert from low carbon back to fossil fuels technologies, pitting short-term political needs against medium-term and long-term energy transition strategies.[61]

7.4 Sustainable Development in the Current EU – Quo Vadis?

This trend of the progressive integration of the rather vaguely defined concept of sustainability with the ephemeral conjunction "sustainable development" in the international and EU policies and law has been observed and heavily discussed during the era of Europe 2020[62] as well as during the current era.[63] Although the conceptual and terminological ambiguity has been repeatedly confirmed,[64] the relationship to the Brundtland Report and Agenda 2030 with SDGs is manifest. At the same time, the EU and Europeans' particularities and the motto "*united in diversity*" need to be appreciated on both the EU level[65] as well as the national level, see, for example, the dramatic differences in SDGs commitments even within one jurisdiction (e.g., various Italian regions).[66]

Arguably, the term "sustainable development" is manipulated and misunderstood and represents an inherent linguistic contradiction.[67] Some suggest that it is doomed due to its lack of pragmatism, disinterest for compromising and balancing and, ultimately, its incapacity to inspire the very needed support across the entire society

[60] Cardesa-Salzmann and Cocciolo (2019), pp. 437–461.

[61] Viñuales (2023), pp. 113–118.

[62] Van Rijswick (2012), pp. 1–6.

[63] Menghwar and Daood (2021), pp. 466–485.

[64] White (2013), pp. 213–217.

[65] Szostek et al. (2022), pp. 231–263.

[66] D'Adamo et al. (2024), p. 5049.

[67] "In its physical aspects, the economy is an open subsystem of the terrestrial ecosystem, which is finite in material resources. As the economic subsystem grows, it incorporates an increasing proportion of the total ecosystem. For this reason, development is not sustainable. The term sustainable development, as it applies to the economy, is a *contradiction*… ." See at Daly (1991), p. 39.

via a multi-stakeholder model. This discourse has been taking place on various platforms in a multi-disciplinary manner.

This has not shaken the firm belief of the UN and the EU, and the literate, as well as contextual and teleological interpretation,[68] of their statements and documents points to the broad concept of sustainable development as launched by the Brundtland Report and entailing as well environmental and social issues.[69] Both the UN and EU keep heralding the idea of a win-win approach to be semi-imposed from above and this generates criticism,[70] labelling such a sustainable development as elusive, contradictory,[71] a cliché[72] and, even more seriously, as undemocratic, not in compliance with fundamental values,[73] and perhaps even not able to stabilize social reproduction, foster global justice and cope with advanced capitalism in the world.[74]

Political science and organizational analysis bring relevant suggestions about adjustments for individuals, institutions, bureaucratic roles and standard operating procedures[75] and it is illustrative to point out the aphorism "where you stand depends upon where you sit."[76] Hence, unlike current EU policies and law, the literature distinguishes more particularities of individual perspectives regarding each pillar of sustainability as projected in sustainable development. Namely, economic sustainable development needs to reconcile not only the interests of environmental and social sustainable development but, as well, needs to balance equity and fiscal soundness,[77] so the burning question is related to conflicts about fairness pragmatism and pragmatic fairness. Social sustainability development is not about meeting the needs of everybody, but about removing barriers blocking individuals from meeting their needs,[78] so the burning question is about the selection of these barriers and the extent of their removal in the context of many issues, such as human rights, gender equity and equality, public participation and the rule of law, promotion of peace and social stability.[79] Environmental sustainability development is perhaps the most controversial and complex in the current literature. The nature and ecosystem should maintain their integrity and the natural capital and natural resources must be harvested no faster than they can be regenerated, while waste must be emitted no

68 Brittain (2016), pp. 134–165.

69 MacGregor Pelikánová and MacGregor (2021), pp. 305–322.

70 Panta (2019), pp. 181–190.

71 O'Riordan (1996).

72 Lélé (1991), pp. 607–621.

73 MacGregor Pelikánová et al. (2024), pp. 1–29.

74 Cardesa-Salzmann and Cocciolo (2019), pp. 437–461.

75 O'Riordan and Jordan (1999), pp. 81–93.

76 Allison (1971).

77 Zhai and Chang (2019), pp. 369–397.

78 Kolk (2016), pp. 23–34.

79 Mensah (2019).

faster than it can be assimilated by the environment.[80] Hence, there the burning question is the very balanced feasibility, as could be implied from Meadows, et al, and his limits. Sustainable development is neither automatic nor obvious. It remains open whether it should be classified as a normal science development or a revolutionary science, aka a paradigm change.[81] The UN and EU keep demanding more actionability of SDGs,[82] while sceptics shower them with criticism[83] and pragmatics point to the impossibility of their satisfaction by 2030. Both proponents and opponents of SDGs explain the difficulty of SDG's satisfaction due to crises and external events, such as the Covid-19 pandemic[84] and the War in Ukraine[85] as well as other events dramatically changing the setting on the macro- as well as micro-level.[86] This is relevant, but still another reason is perhaps even more significant—the generally recognized need of the employment of a multi-stakeholder model is underestimated.

The UN appears slightly ahead, because even the UN Secretary-General António Guterres recognized the issue and summarized the disappointment regarding the lack of progress regarding the meeting of 17 SDGs and their 169 targets in 2030 by stating "Yet today, only 15 per cent of the targets are on track. Many are going in reverse. Instead of leaving no one behind, we risk leaving the SDGs behind".[87] The UN is clearly against any trade-offs or charades regarding SDGs, instead the UN requires maximizing synergies between environmental and social endeavours, in particular climate action and other pro-SDG actions.[88] The UN Under-Secretary-General for Economic and Social Affairs, Li Junhua, summarized it by stating "We must get the SDGs on track and keep the goal of 1.5 degrees alive, an integrated approach that seeks to strengthen synergies between these two global agendas is critical to that end".[89] Indeed, the risks related to climate change and deplorable social situations, as well as other aspects and dimensions reflected by SDGs, are enormous and their extent, nature and quantification exceeds conventional current measurements and models.[90]

Regarding the EU, which, in addition, wants to add integration and welfare state concerns to all environmental and social demands, sustainable development seems to be on even more shaky ground. It is difficult to discuss the process of building bridges between generations when members of current pro-EU groups vigorously protest against the EU perspective about sustainable development. For example,

---

[80] Evers (2018).

[81] Kuhn (1970), p. 92.

[82] Van Tulder et al. (2021), pp. 1–21.

[83] Anderson and Kosnik (2002), pp. 439–475.

[84] Hála et al. (2022), pp. 85–110.

[85] Malý et al. (2023), pp. 119–130.

[86] Cowling and Dvouletý (2023), pp. 402–419.

[87] United Nations (2023b). 18 September 2023.

[88] MacGregor Pelikánová et al. (2024), pp. 1–29.

[89] United Nations (2023a). 13 September 2023.

[90] Rising et al. (2022), pp. 643–651.

pro-EU businesses call for a new deal replacing the European Green Deal and perhaps even the Clean Industrial Deal, pro-EU farmers reject sustainable development in the form of EGD as set by the EU, pro-EU consumers reject sustainable development in the energy choices as set by the EU and pro-EU academics and experts express their very reserved opinions about both the innovation attitude[91] and the innovation implementation by the EU[92] in the (alleged) name of the sustainable development. Many questions emerge, such as whether the current crises (Covid-19 pandemic, War in Ukraine, etc.) threaten steps designed to fight against the climate change and to continue with the energy transition. In searching for possible answers, it is necessary to avoid the black-and-white perspective and superficial short-cuts with long-held assumptions that should no longer be held onto, namely: that fossil fuels are the key to cheap energy, that states face an 'energy trilemma' opposing energy security, affordability and sustainability, and that climate policy is costly and against competitiveness.[93]

Despite all these challenges, possible deficiencies and outdated assumptions, it is obvious that, for decades, the EU institutions, especially the European Commission, have kept on projecting sustainable development into their policy and legislative endeavors and that they are aware that initiatives and following actions regarding sustainable development can (but don't need to) be pro-competitiveness and can be induced from above, but typically they are to be completed from the bottom.[94] Namely, they are to be done by businesses and individuals, either independently or together, see e.g., entities forming the supply chain as an element of sustainable supply chain management.[95] As a matter of fact, the EU relies upon the idea that businesses will engage in three key areas of sustainable development. The areas are green design by products that are more sustainable regarding their composition and packaging, sustainable operations by a more sustainable production, as well as transportation, system and reverse logistics, along with waste management by reuse, recycle and return[96] and this both individually as well as with others while forming sustainable supply chains.[97] In contrast, notorious environmentally polluting and social concerns violating supply chains, such as in the fast fashion industry,[98] are to be rejected by policies, law and even consumers themselves, see the law on corporate sustainability and value chains (CSDDD),[99] the European Green Deal

[91] MacGregor Pelikánová (2019), pp. 13–33.

[92] D'Adamo et al. (2023), pp. 840–852.

[93] Viñuales (2023), pp. 113–118.

[94] MacGregor Pelikánová and MacGregor (2020), pp. 74–108.

[95] Seuring and Müller (2008), pp. 1699–1710.

[96] Hassini et al. (2012), pp. 69–82.

[97] Małys (2023), pp. 255–286.

[98] Niinimäki et al. (2020), pp. 189–200.

[99] Directive (EU) 2024/1760 of the European Parliament and of the Council of 13 June 2024 on corporate sustainability due diligence and amending Directive (EU) 2019/1937 and Regulation (EU) 2023/2859 ("CSDDD" or "CS3D").

and, in particular, the EGD3 New Circular Economy Action Plan (CEAP) and EGD8 Textiles strategy[100] and of course the eagerly expected Clean Industrial Deal. In sum, the materialization, application and enforcement of the outcomes of the new plan for Europe's sustainable prosperity and competitiveness, as declared by the New Political Guidelines,[101] is both eagerly and skeptically expected.

Yes, the 2030 Agenda, as an international law plan for people, planet and prosperity, with 17 SDGs, might appear as revolutionary, yet their key drivers are conventional and traditionalist, see cultural heritage in the context of urban development.[102] Nevertheless, it is a very ambitious plan and its materialization is far from being an easy *fait accompli*[103] and not only the UN appears to be well aware about it. Indeed, the need for innovations, digitalization and circularity for sustainable development is generally recognized[104] and the discussion keeps moving from the "whether" to "how" transition to the modern circular model[105]

The EU refers to Agenda 2030 with SDGs and EU policies overlap with them significantly,[106] but Europeans do not appreciate them fully and are not inspired by them to go for inter- and intragenerational equity. The European Green Deal is the alpha and omega of current EU policies and EU law,[107] and Europeans know it.[108] By the European Green Deal, the environmental sustainable development is anchored at the heart of the modern European integration process[109] and this might be perceived as a "change of paradigm."[110] A successful economic operation in the EU in the third decade of the twenty-first century requires addressing environmental and social aspects in a democratic and collaborative manner,[111] i.e. a legal and profitable conduct in their disregard can be only temporary.

Currently, it is suggested that sustainability is a long-term, if not perpetual, goal desired by (almost) all and sustainable development is a process to reach it via a multi-stakeholder model[112] and they both can flourish only in a respectful, waste-avoiding, bottom-up, collaborative and transparent setting. To build a bridge between generations means knowing what are the needs of these future generations

[100] MacGregor Pelikánová et al. (2024), pp. 1–29.

[101] Von der Leyen U (2024) Political Guidelines for the next European Commission 2024–2029: Europe's Choice, 18 July 2024. Available from: https://commission.europa.eu/document/e6cd4328-673c-4e7a-8683-f63ffb2cf648_en ("New Political Guidelines").

[102] Saleh et al. (2021), pp. 129–156.

[103] Bali and Fan (2019), pp. 96–106.

[104] Skvarciany et al. (2021), pp. 11–34.

[105] Millar et al. (2019), pp. 11–19.

[106] Borchardt et al. (2022).

[107] Schunz (2022).

[108] Kowalska and Bieniek (2022), pp. 607–633.

[109] Bočková (2023), pp. 89–101.

[110] Guégen (2020).

[111] Van Tulder and Van Mil (2023), pp. 1–57.

[112] Panta (2019), pp. 181–190.

and this "knowing" should come based on posed and answered questions and not based on paternalistic assumptions.

Perhaps it would be better to start with a lean version of "smart growth based on European values", which would be well-communicated and endorsed by the majority of Europeans. These Europeans should be able to appreciate and juxtapose economic concerns to environmental and social concerns[113] and benefit by an open, transparent and respectful discussion about setting limits and striking a balance. In the tangible universe, taking more and more, so as to give to more and more recipients means neither sustainability nor development, as a matter of fact this is a self-gratification waste in breach of individual responsibility, respect and creativity, i.e., this violates the values and virtues of European civilization. The other way around, the creativity, individual responsibility, respect and communication, along with the advantages of intangible assets and intellectual property, are the organic elements to build the sustainable development European style. The first step towards this is an honest self-reflection within a genuine multi-stakeholder model—partially done by the UN, but not yet by the European Commission. After all, "Sustainable development is a complex issue, but a simple concept,"[114] so let's be lean, respectful, waste-avoiding and honest, i.e. let's go back to European fundamental values and virtues to ensure success.

Legal Documents

Treaty on European Union, signed at Maastricht on 7 February 1992 (92/C 191/01) ("Treaty of Maastricht")

Treaty of Amsterdam amending the Treaty on European Union, the Treaties establishing the European Communities and certain related acts, signed at Amsterdam on 2 October 1997 (97/C 340/01-05) ("Treaty of Amsterdam")

Lisbon European Council on 23 and 24 March 2000 Presidency Conclusions ("Lisbon Strategy")

Treaty of Lisbon amending the Treaty on European Union and the Treaty establishing the European Community, signed at Lisbon on 13 December 2007 (2007/C 306/01) ("Treaty of Lisbon")

Directive 2014/95/EU of the European Parliament and of the Council of 22 October 2014 amending Directive 2013/34/EU as regards disclosure of non-financial and diversity information by certain large undertakings and groups ("NFRD")

Modern Slavery Act 2015 c.30 from 26 March 2015 (United Kingdom)

[113] MacGregor Pelikánová et al. (2021c), p. 3715.

[114] COM(2019) 22 final Reflection Paper Towards a Sustainable Europe by 2023 from 30 January 2019 ("Reflection Paper").

UN General Assembly Resolution A/RES/71/1 Transforming our world: the 2030 Agenda for Sustainable Development from 25 September 2015 – People, Planet, Prosperity, Peace, and Partnership ("Agenda 2030")

COM(2019) 22 final Reflection Paper Towards a Sustainable Europe by 2023 from 30 January 2019 ("Reflection Paper")

Regulation (EU) 2019/2088 of the European Parliament and of the Council of 27 November 2019 on sustainability-related disclosures in the financial services sector ("SFDR")

Von der Leyen U (2019) Political Guidelines for the next European Commission 2019-2024: A Union that strives for more. My agenda for Europe. Presented to the Members of European Parliament on 16 July 2019. Available from: https://www.europarl.europa.eu/resources/library/media/20190716RES57231/20190716RES57231.pdf ("Political Guidelines")

Regulation (EU) 2020/852 of the European Parliament and of the Council of 18 June 2020 on the establishment of a framework to facilitate sustainable investment, and amending Regulation (EU) 2019/2088 ("Taxonomy Regulation")

Regulation (EU) 2021/1119 of the European Parliament and of the Council of 30 June 2021 establishing the framework for achieving climate neutrality and amending Regulations (EC) No 401/2009 and (EU) 2018/1999 ('European Climate Law')

Corporate Social Reporting Directive Directive (EU) 2022/2464 of the European Parliament and of the Council of 14 December 2022 amending Regulation (EU) No 537/2014, Directive 2004/109/EC, Directive 2006/43/EC and Directive 2013/34/EU, as regards corporate sustainability reporting ("CSRD")

Directive (EU) 2024/1760 of the European Parliament and of the Council of 13 June 2024 on corporate sustainability due diligence and amending Directive (EU) 2019/1937 and Regulation (EU) 2023/2859 ("CSDDD" or "CS3D")

Von der Leyen U (2024) Political Guidelines for the next European Commission 2024-2029: Europe's Choice, 18 July 2024. Available from: https://commission.europa.eu/document/e6cd4328-673c-4e7a-8683-f63ffb2cf648_en ("New Political Guidelines")

References

Allison GT (1971) Essence of decision. Little Brown and Co., Boston

Anderson TK, Kosnik LR (2002) Sustainable skepticism and sustainable development. Case West Reserve Law Rev 53(2):439–475

Balcerzak A, MacGregor RK, MacGregor Pelikánová R, Rogalska E, Szostek D (2023) The EU regulation of sustainable investment: the end of sustainability trade-offs? Entrep Bus Econ Rev 11(1):199–212. https://doi.org/10.15678/EBER.2023.110111

Bali S, Fan YW (2019) Achieving sustainable development goals: predicaments and strategies. Int J Sustain Dev World Ecol 27:96–106. https://doi.org/10.1080/13504509.2019.1692316

Barbiero G (2016) Francis of Assisi and the Wolf: nonviolence as a moral value of biophilia. Vis Sustain 5:28–31. https://doi.org/10.13135/2384-8677/1504

Barbiero G (2021) Affective ecology as development of biophilia hypothesis. In: L'Astorina A, Bergami C, De Lazzari A, Falchetti E (eds) Special issue scientists moving between narratives towards an ecological vision. Visions for Sustainability 16, 5575, 43–78. https://doi.org/10.13135/2384
Basiago AD (1999) Economic, social, and environmental sustainability in development theory and urban planning practice: the environmentalist. Kluwer Academic Publishers, Boston
Bergesen HO (1988) Reformism doomed to failure? A critical look at the strategy promoted by the Brundtland Commission. Int Chall 8(2):6–10
Blewitt J (2017) Understanding sustainable development. Routledge, London. https://doi.org/10.9774/gleaf.9781315465852
Bočková N (2023) The role of the green deal in start-ups' approach to innovation. In: Staníčková M, Melecký L (eds) Proceedings of the 6th International Conference on European Integration 2022. Ostrava: VSB – Technical University of Ostrava (2022), pp 89–101. https://doi.org/10.31490/9788024846057
Borchardt S, Barbero Vignola G, Buscaglia D, Maroni M, Marelli L (2022) Mapping EU policies with the 2030 Agenda and SDGs, EUR 31347 EN, Publications Office of the European Union, Luxembourg, ISBN 978-92-76-60474-7. JRC130904. https://doi.org/10.2760/87754
Brittain S (2016) Justifying the teleological methodology of the European Court of Justice: a rebuttal. Irish Jurist 55:134–165
Cardesa-Salzmann A, Cocciolo E (2019) Global governance, sustainability and the earth system: critical reflections on the role of global law. Transnatl Environ Law 8(3):437–461. https://doi.org/10.1017/S2047102519000098
Chen L (2021) Environmental cost of sustainable development and climate change: can SAARC shift some liability with renewable energy and efficiency? Environ Sci Pollut Res 28(47):67137–67149. https://doi.org/10.1007/s11356-021-15209-2
Cowling M, Dvouletý O (2023) Who is brave enough to start a new business during the Covid-19 pandemic? Balt J Manag 18(3):402–419. https://doi.org/10.1108/BJM-11-2022-0414
Cvik ED, MacGregor Pelikánová R (2021) The significance of CSR during COVID-19 pandemic in the luxury fashion industry – a front-line case study. Eur J Bus Sci Technol 7(1):109–129. https://doi.org/10.11118/ejobsat.2021.005
D'Adamo I, Gastaldi M, Ozturk I (2023) The sustainable development of mobility in the green transition: renewable energy, local industrial chain, and battery recycling. Sustain Dev 31(2):840–852. https://doi.org/10.1002/sd.2424
D'Adamo I, Di Carlo C, Gastaldi M, Rossi EN, Uricchio AF (2024) Economic performance, environmental protection and social progress: a cluster analysis comparison towards sustainable development. Sustainability 16:5049. https://doi.org/10.3390/su16125049
Daly HE (1991) Sustainable development: a contradiction. Desarro Base 15(3):39
De Schutter O (2008) Corporate social responsibility European style. Eur Law J 14(2):203–236. https://doi.org/10.1111/j.1468-0386.2007.00411.x
Dobson A (1996) Environmental sustainabilities: an analysis and a typology. Environ Polit 5:401–428. https://doi.org/10.1080/09644019608414280
Duran DC, Gogan LM, Artene AE, Duran V (2015) The components of sustainable development - a possible approach. Procedia Econ Finance 26:806–811. https://doi.org/10.1016/S2212-5671(15)00849-7
Evers BA (2018) Why adopt the sustainable development goals? The case of multinationals in the Colombian coffee and extractive sector: Master Thesis Erasmus. University Rotterdam, Rotterdam
Feil A, Schreiber D (2017) Sustainability and sustainable development: unraveling overlays and scope of their meanings. Cadernos 14(3):667–681. https://doi.org/10.1590/1679-395157473
Fitzpatrick H (2023) A review of worldviews beyond sustainability. Potential avenues for human-nature connectedness. Vis Sustain 19(7309):9–57. https://doi.org/10.13135/2384-8677/7309
Grabara J (2019) Sustainable development - never fulfilled dream. Quality – Access to Success 20(1):565–570

Guégen D (2020) Green deal: the long battles lying ahead. EurActiv. https://www.euractiv.com/section/energy-environment/opinion/green-deal-the-long-battles-lying-ahead/. Accessed 8 Nov 2023

Hála M, Cvik ED, MacGregor Pelikánová RM (2022) Logistic regression of Czech luxury fashion purchasing habits during the Covid-19 pandemic – old for loyalty and young for sustainability? Folia Oeconomica Stetinensia 22(1):85–110. https://doi.org/10.2478/foli-2022-0005

Harlow J, Golub A, Allenby B (2013) A review of utopian themes in sustainable development discourse. Sustain Dev 21(4):270–280. https://doi.org/10.1002/sd.522

Harwood RR (1990) The history of sustainable agriculture. In: Edwards CA et al (eds) Sustainable farming systems, pp 3–19

Hassini E, Surti C, Searcy C (2012) A literature review and a case study of sustainable supply chains with a focus on metrics. Int J Prod Econ 140(1):69–82. https://doi.org/10.1016/j.ijpe.2012.01.042

Hector DC, Christensen CB, Petrie J (2014) Sustainability and sustainable development: philosophical distinctions and practical implications. Environ Values 23(1):7–28. https://doi.org/10.3197/096327114X13851122268963

Holden E, Linnerud K, Banister D (2014) Sustainable development: our common future revisited. Global Environ Change 26:130–139. https://doi.org/10.1016/j.gloenvcha.2014.04.006

Huttmanova E, Valentiny T (2019) Assessment of the economic pillar and environmental pillar of sustainable development in the European Union. Eur J Sustain Dev 8(2):289–298. https://doi.org/10.14207/ejsd.2019.v8n2p289

Keeble BR (1988) The Brundtland Report: 'Our Common Future'. Med War 4(1):17–25. https://www.jstor.org/stable/45353161

Klarin T (2018) The concept of sustainable development: from its beginning to the contemporary issues. Zagreb Int Rev Econ Bus 21(1):67–94. https://doi.org/10.2478/zireb-2018-0005

Kolk A (2016) The social responsibility of international business: from ethics and the environment to CSR and sustainable development. J World Bus 51(1):23–34. https://doi.org/10.1016/j.jwb.2015.08.010

Kowalska A, Bieniek M (2022) Meeting the European green deal objective of expanding organic farming. Equilibrium 17(3):607–633. https://doi.org/10.24136/eq.2022.021

Kuhn TS (1970) The structure of scientific revolutions. University of Chicago Press, Chicago, p 92

Lauesen LM (2013) OECD principles of corporate governance and CSR. In: Idowu SO, Capaldi N, Zu L, Gupta AD (eds) Encyclopedia of corporate social responsibility. Springer, Berlin, Heidelberg. https://doi.org/10.1007/978-3-642-28036-8_25

Leczykiewicz D (2005) Common commercial policy: the expanding competence of the European Union in the area of international trade. German Law J 6(11):1673–1685. https://doi.org/10.1017/S2071832200014607

Lélé SM (1991) Sustainable development: a critical review. World Dev 19(6):607–621. https://doi.org/10.1016/0305-750X(91)90197-P

MacGregor Pelikánová R (2017) Constantine's Christianity for the (dis)integrated EU - Déjà vu of Constantine's Roman governance reflecting of the mistrial of Jesus for EU? Dialogo 4(1):81–98. https://doi.org/10.18638/dialogo.2017.4.1

MacGregor Pelikánová R (2019) R&D expenditure and innovation in the EU and selected member states. JEMI – J Entrep Manag Innov 15(1):13–33. https://doi.org/10.7341/20191511

MacGregor Pelikánová R (2024) EU legislative proposals to involve customers in the greening of the economy. Lawyer Q 2:218–235. https://tlq.ilaw.cas.cz/index.php/tlq/article/view/597

MacGregor Pelikánová R, Hála M (2021) CSR unconscious consumption by Generation Z in the COVID-19 era – responsible heretics not paying CSR bonus? J Risk Financ Manag 14(8):390. https://doi.org/10.3390/jrfm14080390

MacGregor Pelikánová R, MacGregor RK (2020) The EU puzzling CSR regime and the confused perception by ambassadors of luxury fashion businesses: a case study from Pařížská. Central Eur Bus Rev 9(3):74–108. https://doi.org/10.18267/j.cebr.240

MacGregor Pelikánová R, MacGregor RK (2021) The Covid-19 as interpretation instrument for the content of corporate social responsibility and its reporting in the EU. Lawyer Q 2:305–322
MacGregor Pelikánová R, Rubáček F (2022) Taxonomy for transparency in non-financial statements – clear duty with unclear sanction. Danube 13(3):173–195. https://doi.org/10.2478/danb-2022-0011
MacGregor Pelikánová R, Sani M (2023) Luxury, slow and fast fashion – a case study on the (un)sustainable creating of shared values. Equilibrium. Q J Econ Econ Policy 18(3):813–851. https://doi.org/10.24136/eq.2023.026
MacGregor Pelikánová R, Cvik ED, MacGregor RK (2021a) Addressing the COVID-19 challenges by SMEs in the hotel industry – a Czech sustainability message for emerging economies. J Entrep Emerg Econ 3(4):525–546. https://doi.org/10.1108/JEEE-07-2020-0245
MacGregor Pelikánová R, MacGregor RK, Černek M (2021b) New trends in codes of ethics: Czech business ethics preferences by the dawn of COVID-19. Oeconomia Copernicana 12(4): 973–1009. https://doi.org/10.24136/oc.2021.032
MacGregor Pelikánová R, Němečková T, MacGregor RK (2021c) CSR statements in international and Czech luxury fashion industry at the onset and during the COVID-19 pandemic—slowing down the fast fashion business? Sustainability 13(7):3715. https://doi.org/10.3390/su13073715
MacGregor Pelikánová R, Sani M, Rubáček F (2024) Sustainable and responsible creation of shared values in the fast fashion industry – I will believe it when I see it. Vis Sustain 21(8873):1–29. https://doi.org/10.13135/2384-8677/8873
Malý M, Cvik ED, MacGregor Pelikánová R (2023) EU sanctions against the Russian Federation and their implications for the Foreign Trade of the Czech Republic. AGRIS On-line Pap Econ Inform 15(3):119–130. https://doi.org/10.7160/aol.2023.150310
Małys Ł (2023) The approach to supply chain cooperation in the implementation of sustainable development initiatives and company's economic performance. Equilibrium. Q J Econ Econ Policy 18(1):255–286. https://doi.org/10.24136/eq.2023.008
Menghwar PS, Daood A (2021) Creating shared value: a systematic review, synthesis and integrative perspective. Int J Manag Rev 23(4):466–485. https://doi.org/10.1111/ijmr.12252
Mensah J (2019) Sustainable development: meaning, history, principles, pillars, and implications for human action: literature review. Cogent Soc Sci 5(1). https://doi.org/10.1080/23311886.2019.1653531
Millar N, McLaughlin E, Börger T (2019) The circular economy: swings and roundabouts? Ecol Econ 158:11–19. https://doi.org/10.1016/j.ecolecon.2018.12.012
Niinimäki K et al (2020) The environmental price of fast fashion. Nat Rev Earth Environ 1:189–200. https://doi.org/10.1038/s43017-020-0039-9
O'Riordan T (1996) Environmentalism on the move. In: Douglas I, Huggett RJ, Robinson M (eds) Companion encyclopaedia of geography. Routledge, London
O'Riordan T, Jordan A (1999) Institutions, climate change and cultural theory: towards a common analytical framework. Global Environ Change 9:81–93
Panta ND (2019) Clashing perspectives on sustainable development. Stud Bus Econ 14(1): 181–190. https://doi.org/10.2478/sbe-2019-0014
Purvis B, Mao Y, Robinson D (2019) Three pillars of sustainability: in search of conceptual origins. Sustain Sci 14(3):681–695. https://doi.org/10.1007/s11625-018-0627-5
Rendtorff JD (2019) Sustainable development goals and progressive business models for economic transformation. Local Econ 34(6):510–524. https://doi.org/10.1177/0269094219882270
Rising J, Tedesco M, Piontek F et al (2022) The missing risks of climate change. Nature 610:643–651. https://doi.org/10.1038/s41586-022-05243-6
Rubáček F, MacGregor Pelikánová R, MacGregor RK (2023) The sustainability of ESAs triumvirate for sustainability-related disclosures in the financial sector – all for one and one for all. Acta Universitatis Lodziensis. Folia Iuridica 105:123–143. https://doi.org/10.18778/0208-6069.105.08

Saleh Y, Mahat H, Hashim M, Nayan N, Suhaily S, Ghazali MKA, Hayati R, Utami RKS (2021) The development of a construct in the heritage urban sustainability index. Vis Sustain 17(6082): 129–156. https://doi.org/10.13135/2384

Schunz S (2022) The 'European Green Deal' – a paradigm shift? Transformations in the European Union's sustainability meta-discourse. Polit Res Exchange 4(1). https://doi.org/10.1080/2474736X.2022.2085121

Seuring S, Müller M (2008) From a literature review to a conceptual framework for sustainable supply chain management. J Clean Prod 16(15):1699–1710. https://doi.org/10.1016/j.jclepro.2008.04.020

Skvarciany V, Lapinskaite I, Volskyte G (2021) Circular economy as assistance for sustainable development in OECD countries. Oeconomia Copernicana 12(1):11–34. https://doi.org/10.24136/oc.2021.001

Stark R (2009) God's Battalions. The case for the Crusades. Harper One, New York, pp 67–70

Stark R (2014) How the West Won. ISI Books, Delaware

Sterling S (2010) Learning for resilience, or the resilient learner? Towards a necessary reconciliation in a paradigm of sustainable education. Environ Educ Res 16:511–528

Szostek D, Balcerzak A, Rogalska E, MacGregor Pelikánová R (2022) Personality traits and counterproductive work behaviors: the moderating role of demographic characteristics. Econ Sociol 15(4):231–263. https://doi.org/10.14254./2071-789X.2022/15-4/12

United Nations (2023a) Press Release. Governments must seek win-win synergies by tackling climate and sustainable development crises together, urges expert group report. 13 September 2023. Available from: https://www.un.org/sustainabledevelopment/blog/2023/09/press-release-governments-must-seek-win-win-synergies-by-tackling-climate-and-sustainable-development-crises-together-urges-expert-group-report/. Accessed 18 Nov 2023

United Nations (2023b) Press Release. Responding to cascading global crises, SDG Summit launches new phase of accelerated action on the Sustainable Development Goals. 18 September 2023. Available from: https://www.un.org/sustainabledevelopment/blog/2023/09/press-release-responding-to-cascading-global-crises-sdg-summit-launches-new-phase-of-accelerated-action-on-the-sustainable-development-goals/. Accessed 18 Nov 2023

Van Rijswick HFMW (2012) The road to sustainability: how environmental law can deal with complexity and flexibility. Utrecht Law Rev 8(3):1–6

Van Tulder R, Van Mil E (2023) Principles of sustainable business. Frameworks for corporate action on the SDGs. Routledge, London and New York

Van Tulder R, Rodrigues SB, Mirza H et al (2021) The UN's Sustainable Development Goals: can multinational enterprises lead the decade of action? J Internal Bus Policy 4:1–21. https://doi.org/10.1057/s42214-020-00095-1

Viñuales JE (2013) The rise and fall of sustainable development. Rev Eur Comp Int Environ Law. Special Issue: Assessing Progress in International Environmental Law 22(1):3–13. https://doi.org/10.1111/reel.12021

Viñuales JE (2023) The Ukraine War and the energy transition. RED 1(5):113–118. https://doi.org/10.3917/red.005.0123

White MA (2013) Sustainability: I know it when I see it. Ecol Econ 86:213–217. https://doi.org/10.1016/j.ecolecon.2012.12.020

Zhai T, Chang Y (2019) Standing of environmental public-interest litigants in China: evolution, obstacles and solutions. J Environ Law 30:369–397. https://doi.org/10.1093/jel/eqy011

Chapter 8
Engagement of Private Subjects with Sustainability: Corporate Social Responsibility

The term "corporate social responsibility" (CSR) is understood as a modern label of a universally and generally applied self-regulating business model that helps a business (entity) to be socially accountable to itself, its stakeholders and even the public-at-large. However, even a cursory observation reveals a set of ambiguities and contradictions. CSR is neither new nor "corporate" specific, nor necessarily self-imposed nor only accountability based. Legal entities engaging in business have been a fact of life of European civilization for a millennia, and the issue of their accountability v responsibility v liability has accompanied them through, practically, their entire history. Consequently, CSR has important historical and conceptual roots (Sect. 8.1) which need to be explored and appreciated before the engagement with the modern understanding of CSR launched by Bowen and visualized by Carroll (Sect. 8.2). However, CSR is very dynamic and so the static visualization needs to be complemented by the dynamics of CSR stages, phases and versions (Sect. 8.3) so as to appreciate what is the status quo of CSR (Sect. 8.4).

8.1 History and Conceptual Roots of Corporate Social Responsibility

In general, CSR is presented as the demand upon businesses to meet society's expectations in their practice.[1] Nowadays, CSR is often used vis-à-vis all types of businesses, other legal entities and even individuals, and the exponential growing legislation often turns the responsibility into a liability, while dealing with much more than only social matters in the narrow sense. Despite all that, the term CSR is firmly established and used without serious complaints. It almost appears that CSR has a life of its own, detached from its very wording and its own history. Well, a

[1] Gössling and Vocht (2007), pp. 363–372.

R. MacGregor Pelikánová, *Sustainability in Europe*,
https://doi.org/10.1007/978-3-031-74689-5_8

deeper analysis of the historical and conceptual roots of CSR reveals that "corporate" and "social" and "responsibility" are definitely appropriate and relevant terms and that CSR is an organic product of a development led by heterogenous groups.

Starting with "corporation", it is necessary to point out that the foundation of almost (if not all) business forms used currently in Europe was already laid down by Roman law, while the word "corpus" means, in Latin, a body or a body of people. In the fifth century, during the massive (re)codification of Roman law, *Corpus Iuris Civilis*,[2] orchestrated by the Byzantine Emperor Justinian and his illustrious wife Theodora, a set of such business forms was well recognized, see *universitas*, *corpus* or *collegium*. Typically, these bodies needed to be approved by a state authority, had the legal capacity to own and dispose with property as well as to sue and to be sued and to act through representatives. Exactly all this was rediscovered in North Italy in the eleventh century and became a popular subject for a myriad of famous Italian jurists becoming glossers and commentators. Arguably, the first commercial entities matching up with our current understanding of the form of business corporations and companies emerged in the fourteenth century, see *City of London Corporation* in London, *Stora Kopparberg* in Falun in Sweden, *Barovier & Toso* in Venice, *Torrini Firenze* in Florence, etc. Further, in medieval times, traders, merchants and artisans often used the form of partnership and group in associations called guilds. In the seventeenth century, the era of Hansa and guilds were over, while overseas discoveries and colonization, along with international trade, was booming and thus creating conditions for the establishment of more capital than the personal type of companies, see trading companies such as the Dutch West India Company. The potential for risk and both dramatic profits or losses of everything increased dramatically, while the regulation of insolvency envisaged already by the Code of Hammurabi (eighteenth century BC),[3] the Twelve Tables of the Roman Republic (450 BC)[4] and the *Corpus Iuris Civilis* (534) reached a new stage.

In the sixteenth century, probably the first bankruptcy legislation was enacted during the 2nd session of the 8th Parliament of Henry VIII and via the Statute of Bankrupts, aka *An Acte againste suche persones as doo make Bankrupte* from 1542,[5] which was highly suspicious regarding insolvent persons and had for its aim the prevention of their irresponsible and fraudulent behaviour. Namely, the Lord Chancellor was given the power to sell off the property of debtors and even imprison debtors not willing to pay their debts and, at the same time, wanting to flee and/or keep their houses and "at their own will and pleasure consume the substance obtained by credit of other men, for their own pleasure and delicate living, against all reason, equity and good conscience." Although the original "debt slavery"

[2] *Corpus Iuris Civilis (Codex, Digest or Pandects, Institutes/Institutiones+ Novellae Constitutiones)*, 529–534.

[3] Code of Hammurabi, 1755 - 1750 B.C., a collection of 282 rules.

[4] *Lex Duodecim Tabularum – Duodecim Tabulae* (Law of the Twelve Tables), 449 BC.

[5] *An Acte againste suche persones as doo make Bankrupte* 34 & 35, 8th Parliament of Henry VIII, c. 4 of 1542 ("Statute of Bankrupts of 1542").

approach was appeased by the medieval canon law allowing the debtor to have a fresh start after ceding to his creditors all his goods, the Statute of Bankrupts 1542 still relied on the use of debtor's prisons and their abolishment occurred only in the nineteenth century. However, this reliance was not absolute and the Statute of Bankrupts 1542 offered several alternatives and so the debtor did not need to end up in jail but instead the Lord Chancellor might discharge him, provided he disclosed all his assets and his creditors were satisfied from them proportionally, i.e. based on the *pari passu* principle. To put it differently, the debtor was expected to be responsible and provide honest, genuine information about his assets and general business standing—this might be perceived as a precursor of the current CSR reporting.

In the eighteenth century, the foundations of modern political economy were laid down and, conventionally, Adam Smith is presented as the father of economics and/or the father of capitalism. However, his massive contribution is often misleadingly simplified in the concept of a totally free hand (the famous 'Invisible Hand') of a market and he is ultimately presented as the advocate of ruthless individualism and employment of businesses forms, such as corporations, for profit maximization "at any price." Both of his key works, The Theory of Moral Sentiments (1759) and An Inquiry into the Nature and Causes of the Wealth of Nations (1776) entail responsibility concerns, even moral philosophy aspects. Adam Smith recognized, as sources of a long term prosperity, the division of labour in a free market, while he highlighted the role of state and government to provide an enforceable legal framework and the protection of intellectual property and the role of businesses and their capacity to engage with ethics and charity.

The nineteenth century was not only the century of the industrial revolution but as well the century of the systemic recognition of the responsibility of businesses vis-à-vis the entire society, i.e. the idea of reckless businesses expected to observe only strict law limits and mandatory set law duties is wrong. As a matter of fact, the industrial revolution entailed not only the massive industrial development facilitated by numerous inventions and innovations, but, as well, drastic social and environmental changes which made English poet William Blake speak about the infamous "dark satanic mills."[6] Individual entrepreneurs such as Robert Owen and George and Richard Cadbury engaged in various endeavours to improve working conditions and suggested that the welfare of employees is a consideration and can ultimately create a win-win solution.[7] Exactly this was at the centre of the Rhenish model of the social market with its insurance funds and leading to German company social welfare programs focused not only on health insurance, housing, and company stores[8] but as well on nation-building.[9] Ultimately, these social welfare programs demonstrated social responsibility, attempted to "deflect criticism of industrial capitalism" as too

[6]Blake (1804/1970).

[7]Husted (2015), pp. 125–141.

[8]Hilger (1998), pp. 50–76.

[9]Spencer (1979), pp. 40–64.

focused on the "reckless pursuit of profit"[10] and inspired Bismarck's social welfare plan,[11] which consisted of: the German Health Insurance Act 1883 establishing the oldest national social health insurance system,[12] the German Accident Insurance Act 1884 addressing the insurance of workers,[13] the Limitation of Working Hours Act 1887 putting limits on the labour of women and children and shortening working hours[14] and the German Old Age and Invalidity Insurance Act of 1889.[15] Bismarck wanted to convince workers that they are better off with his government than with Socialists, but his combination of coercion, charity and implied legislation did not achieve it in the manner as he wanted.[16]

In addition to these social concerns, businesses and individuals alike started to recognize that businesses are implicated in the degradation of the environment and that the increasing emissions are bad and that the reduction of waste increases the value of inputs, as suggested by the English polymath Charles Babage. However, it needs to be admitted that, in the nineteenth century, there clearly prevailed the social over environmental protection dimension and, especially in Germany, the proactive participation of businesses in innovative social welfare programs can be contrasted with their limited stance with environment protection. For example, the German smoke emissions regulation in nineteenth century,[17] was predominantly ignored by the business community.[18]

Regarding various levels of awareness and commitment maturity in the nineteenth century, the scientific management expert, Henry Gantt, put all of this together and implied, from these "dark satanic mills", and various environmental protection and welfare endeavours, a systemic business responsibility in 1919: "We all realize that any reward or profit that a business arbitrarily takes, over and above that to which it is justly entitled for services rendered, is just as much the exercise of autocratic power and a menace to the industrial peace of the world, as the autocratic military power of the Kaiser was a menace to international peace. This applies to Bolshevists as well as to Bankers."[19] This strong, rather political science and public

[10] Husted (2015), pp. 125–141.

[11] Hilger (1998), pp. 50–76.

[12] *Gesetz, betreffend die Krankenversicherung der Arbeiter (KVG) vom 15.Juni 1883 - Deutsches Reichsgesetzblatt Band 1883, Nr. 9, Seite 73–104* ("German Health Insurance Act 1883").

[13] *Unfallversicherungsgesetz vom 6.Juli 1884 - Deutsches Reichsgesetzblatt Band 1884, Nr. 19, Seite 69* ("German Accident Insurance Act 1884").

[14] *Einschränkung der Frauenarbeit und Kinderarbeit Gesetz von 17.Juni 1887 Deutsches Reichsgesetzblatt Band 1887, Nr. 88, Seite 913* ("Limitation of Working Hours Act 1887").

[15] *Gesetz betreffend die Invaliditäts- und Altersversicherung vom 22.Juni 1889 - Deutsches Reichsgesetzblatt Band 1889, Nr. 13, Seite 97* ("German Old Age and Invalidity Insurance Act 1889").

[16] Friedrich (1996), p. 222.

[17] *Dampfkessel-Gesetzgebung in Preußen von 1831, Preußische Gewerbeordnung von 1845, Reichsgewerbeordnung von 1871*, etc.

[18] Uekoetter (1999), pp. 641–676.

[19] Gantt (1919), pp. 7–8.

law language was further transformed in his very clear conceptual statement with clear ramifications into the private sphere, i.e. "The business system must accept its social responsibility and devote itself primarily to service, or the community will ultimately make the attempt to take it over in order to operate it in its own interest."[20]

Sadly, this is often overlooked and it cannot be overemphasized that CSR is not just a mere product of endeavours in the USA in the context of the antitrust battle, the Great Depression and post-Second World War issues. Instead, CSR is a practice from the nineteenth century which developed in parallel in several countries, such as in the United Kingdom, the United States, Japan, India, and Germany.[21] And in each of these jurisdictions there were sets of struggles marked by various levels of maturity vis-à-vis different dimensions and aspects of the responsibility of businesses and by cultural and historic particularities. However, in each and every jurisdiction, there was discussed the ultimate motivation, while suggesting profit-orientation, humanitarianism or their mix.[22] This goes back to the eternal juxtaposition of the deontological approach and consequentialist approach and Stark correctly pointed to the complex business reality and inherently mixed motives linked to business ethics.[23] Indeed, the modern CSR has a rather surprising history and heterogenous roots. It is undeniable that it is the outcome of the creative endeavours of many individuals having their own vision of responsibility and society and ready to look for alternatives and options about how to resolve economic, environmental and social problems.[24] Certainly, pre-CSR and early CSR endeavours in the nineteenth century demonstrated a paternalistic approach, thus tainting our perception of their version of the welfare- and environmentally aware capitalism. However, at the same time they demonstrated a fresh, open-minded and particular approach. In contrast, modern CSR attempts to go for the EU (if not global) harmonization, standardization and perhaps even unification, and maybe this departure from its origins might be too ambitious … or not, let see how Bowen and Carroll proposed to solve this issue.

8.2 Modern CSR Envisaged by Bowen and Visualized by Carroll

As already mentioned, the idea of tying the power to the responsibility, vis-à-vis subjects, and of making businesses responsible for the economic, environmental and social impacts of their business conduct, emerged in the nineteenth century predominantly in the United Kingdom, but promptly spread in countries strongly impacted

[20] Gantt (1919), p. 15.

[21] Husted (2015), pp. 125–141.

[22] McCreary (1968), pp. 24–49.

[23] Stark (1993), pp. 38–48.

[24] Husted (2015), pp. 125–141.

by the industrial revolution. Way before the literature and legislatures came up with the idea to impose an extended version of the reasonable care duty of businesses, i.e. to make businesses responsible for their impact on the society in the aftermath of the Second World War, several foundations were suggested, such as the idea of stewardship, the idea of trusteeship, the principle of the common good, etc. However, these suggestions were subject to vivid theory and conceptual disputes, demonstrating a strong tension between social Darwinism, on the one hand, and welfare capitalism or Andrew Carnegie's Gospel of Wealth, on the other. The core and focal point of this dispute clearly moved to the US, where many businesses were rather reluctant to participate in social and environmental reforms. They deeply believed that "smoke means prosperity" and corporate welfare is controversial. Given the predominant *laissez-faire* approach of the times, they believed that it is absurd to entitle a man to a living wage and that taking from the strong to give to the weak means encouraging weakness.[25]

However, several events have shaken this scepticism and passivity and forced a change of attitude. The Sherman Antitrust Act of 1890, the Clayton Act of 1911, the Federal Trade Commission Act of 1911 and their strong case law made it clear that the protection of free and unfettered competition is a genuine and enforced target of the public law. The growing concern to protect the process of competition in the largest sense implies the reduction of the freedom of businesses to do "whatsoever." Similarly, external international and political events, such as wars, as well as economic and other crises, such as the Great Depression, have forced businesses to depart from their paternalistic and/or passive position vis-à-vis their societal responsibility. Challenges and opportunities are brought by crises,[26] while fundamental law doctrines such as the rule of law and the focus on democratic accountability and transparency progressively extend as well to private law subjects. The Harvard Law Review became a platform for the discussion about the proper purpose of companies, corporations and even businesses in general, while Adolph A. Berle argued for profit generation for shareholders and Merrick Dodd argued for a broader pool of subjects since economic institutions should serve both shareholders and society and shareholders.[27]

During the Second World War and after, the trend was to get everyone involved, especially those with resources, regardless of their constitutional position. This call received dramatic contours around 1950 when the cold war progressed, the trial with the Rosenbergs took place and the post-war boom, aka economic expansion, led to the golden age. The economic power of businesses kept growing, as well as the general call for the rule of law and the democratic accountability of all and, after 1945, expectations increased. The US Gross Domestic Product increased exponentially, the US economy was moving towards representing 1/3rd of the total of the world's industrial output, the US companies behind it were becoming super

[25]Husted (2015), pp. 125–141.

[26]D'Adamo and Lupi (2021), p. 1861.

[27]Roszkowska-Menkes (2021).

powerful and the information about it went world-wide. The writing was on the wall and it was just a question of time when somebody would cross the Rubicon and tie both ends—power and responsibility for all of its consequences—together. This someone was a professor, Howard R. Bowen, who was involved in both public service and academia and who was deeply convinced about the inevitable ethical dimension of business conduct. Older and/or senior faculty at the University of Illinois did not like his curriculum on the social responsibility of businessmen, they accused him of being anti-business and ultimately they forced him to leave in 1950 ... and then he wrote his milestone book, thus launching the modern CSR and directly leading to our current perception that denying CSR is anti-business.

In 1953, Howard R. Bowen published his highly influential book, "Social Responsibilities of the Businessman", in which he explained the relationship between power and responsibility in the business setting and, in particular, he described the power and impact of the largest US businesses on the entire society.[28] The exercise of such power should be neither totally discretionary nor without any responsibility for its consequences—such a business has to carry the CSR. This idea was not contradictory to the emerging thoughts in the UN and even more the OECD, which had been issuing recommendations for economic development and even sustainable economic growth, corporate governance (CG) and even CSR.[29] As a matter of fact, the OECD Guidelines for Multinational Enterprises could be perceived as a comprehensive code of conduct, with recommendations on responsible business conduct addressed by governments to multinational enterprises operating in or from over 50 countries, including EU member states. Already in its original version from 1976 it envisaged CSR, underwent numerous revisions in 1979, 1982, 1984, 1991, 2000, 2011, and 2023 ("OECD Guidelines 2023),[30] and currently attempts to encourage the positive contributions of multinational enterprises "to economic, environmental, and social progress and to minimise the adverse impacts on the matters that may be associated with an enterprise's operations, products and services. Responsible business conduct can enable the creation of a level playing field across global markets, foster a dynamic and well-functioning business sector, and enhance the business contribution to sustainable development outcomes, including solutions to address and respond to climate change." (Preface p. 10).

Ever since 1953, the literature has kept progressively moving to reinforce this imposition of duties related to sustainable development, not only upon public, but as well private, subjects. This academic trend became boosted by the progressive admission of the objective shrinking (or at least not growing) of tangible resources, see The Limits to Growth by Meadows et al. from 1972 and the UN Brundtland

[28]Bowen (2013).

[29]Lauesen (2013), pp. 1771–1777.

[30]OECD (2023), OECD Guidelines for Multinational Enterprises on Responsible Business Conduct, 8 June, 2023, OECD Publishing, Paris. https://doi.org/10.1787/81f92357-en ("OECD Guidelines 2023").

Report from 1987.[31] Nevertheless, the pragmatism and scepticism regarding either massive or openly altruistic CSR kept playing a pivotal role, see Free to Choose: A Personal Statement, by Milton Friedman and his wife Rose Friedman from 1980, which argues that the free-market operation is the best choice because it works for the best for all members of society and brings a general prosperity. The resulting shareholder theory is based upon the protection of personal freedoms (of investors) and economic efficiency and, similar to propositions advanced by Bowen, Meadows and Brundtland, wants to achieve sustainability by avoiding waste and, differently from them, assign the decision about it. Namely, Friedman rejected interventionist government policies, opposed Keynesian interventionalism via various policies and advocated for the state minimalist approach. Consequently, Friedman rejected the delegation of cost decisions and demanded that businesses do legally all possible to maximize profits of the shareholders.[32] Later, this approach of the shareholder theory was summarized via the famous and partially misleading motto "what is good for business is good for society."[33] The public sustainability concerns vested in such a private social responsibility means CSR, which is rather elementary and defensive. The Friedman's shareholder theory is pro-CSR, provided that it creates a value for shareholders, e.g. sustainable expense is acceptable if it avoids a fine imposed by the state but not if it supports some ephemeral pro-bono projects.

The classic confrontation between Keynes' view and the Friedman's view about sustainability and CSR could not be fully developed because the scenery for such an eloquent and sophisticated dispute between Keynes' and the Friedman's supporters and the considerations of what was sufficiently good to increase shareholder value dramatically changed in the 1980s, particularly due to the economic growth slowdown, stagflation and the increase in the population, especially in the Third World. The more and more globalized world witnessed the pressures of both poverty and the increasing population, making it extremely difficult for both developed and developing countries to behave pro-socially and pro-environmentally. Both the interventionism and defensive CSR started to be unsustainable in the interconnected world, with mushrooming unsustainable practices. Sustainability could not be taken care of only by states and shareholders. Friedman's shareholder theory lifted the veil and was proclaimed, whether in the form of a company or a corporation, to be a mere vehicle for the business conduct of its shareholders, but manifestly as well other (passive) investors needed to be considered and involved.

The traditionalist shareholder theory was paralleled by the stakeholder theory enlarging the circle of beneficiaries while referring to the same fundamental values, especially the autonomy and free will, and adding concerns regarding interconnected relationships between a business and its customers, suppliers, employees, investors, communities, etc.—its stakeholders, see *Strategic Management: a Stakeholder*

---

[31] UN Annex to document A/42/427 Report of the World Commission on Environment and Development: Our Common Future from 4 August 1987 ("Brundtland Report").

[32] Friedman (2007), pp. 173–178.

[33] Porter and Ignatius (n.d.).

Approach by R. Edward Freeman from 1984.[34] Pursuant to this stakeholder theory, the chairperson, typically a CEO, was perceived as a direct employee of the business, and as an indirect employee of the stakeholders (not only the shareholders) with the job to maximize their profits. Addressing social or environmental needs by such a business/CEO without aiming at increasing profits and/or benefits is, from this perspective, a waste of stakeholder resources, perhaps even a misappropriation of the entrusted funds and functions.

The concept of sustainability is rejected neither by the traditionalist shareholder theory nor by the stakeholder theory, but the business management decisions about sustainability are to be judged from a shareholder's or a stakeholder's perspective, because the freedom of will and protection of private property are fundamental values. Following the Bowen logic that he who impacts has the responsibility for it, the question is whether we consider only the most (financially) impacted or whether we consider a broader pool of impactees. In sum, the business does not need to work for all stakeholders and/or stakeholders (this would not be feasible), but must take them into consideration. It is about balancing the various interests of various subjects and ultimately modifies the business's ability to generate profits. This was perceived as both trade-offs and limits until the mutual win-win CSV theory was presented.

However, it would be superficial to believe that these two US origin theories led to two models in Europe. As a matter of fact, many modified versions influenced by local particularities emerged in Europe in Bowen's aftermath. For example the famous pro-sustainability Nordic governance model combined social liberalism, majority shareholder power, a concentrated state and the counterbalancing of family and foundation shareholdings by strong minority shareholder protection.[35] This Nordic governance model was conceived as "outside-in" because non-shareholder stakeholders should be considered even before the shareholders, plus the minority shareholders and employees should be instrumental for sharing values.[36] Consequently, this model was labelled as implicit CSR and was perceived as a result of a consensual political culture, welfare state and business-government-labour union partnership.[37] Such a Nordic model was strongly applied until 1990 in Denmark, Finland, Sweden and even Norway, where nevertheless, the setting was slightly different due to the natural oil reserve situation. Simultaneously, a similar model emerged in Germany and was projected even into the Corporate law. Hence, the decision making by businesses was not entirely in the hands of the majority shareholders and therefore the idea of a strict commitment to maximize profit for majority shareholders could not exist in the crystalized form.

In addition, regardless of which theory is endorsed, it must be realized that the freedom of will and the protection of private property occurs in the society with

[34] Freeman (1984).

[35] Mähönen and Johnsen (2019), pp. 218–231.

[36] Carson et al. (2015), pp. 17–19.

[37] Carson et al. (2015), p. 21.

interlinking and intra-linking, i.e. businesses do not act alone in a vacuum and individual responsibility is the foundation of our civilization. Although each of the above-mentioned theories could even currently serve as a valid normative theory, it appears that, not only the shareholder theory, but as well the stakeholder theory have both passed beyond their golden age.[38] Namely, the dogma about generating profits (not only financial) is progressively overshadowed by the dogma about creating value, and business conduct is assessed while considering more than merely the economic dimension. Several dimensions of a business's conduct lead to several consequences and they mean several responsibilities.

In 1991, Archie B. Carroll consolidated this idea by tying together power and consequences and visualized it via the famous four-layers pyramid with one responsibility on each layer (economic, legal, ethical and discretionary (philanthropic) responsibility).[39] Such a consolidation was the result of over ten years of endeavours, because Carroll suggested, already in 1979, that "For a definition of social responsibility to fully address the entire range of obligations business has to society, it must embody the economic, legal, ethical, and discretionary categories of business performance. These four basic expectations reflect a view of social responsibility ..."[40] In doing so, Carroll considered Friedman's profit making only, Davis's going beyond profit making, Manne's voluntary activities, Ackerman's giving way to social responsiveness, etc. Namely, already by 1979 Carroll was very close to reaching a summary and, on his way to visualizing, he could take advantage of a more symbolic description of CSR by Peter F. Drucker, expressed in 1984, by his stating, "The proper social responsibility of business is to tame the dragon, that is to turn all social problems into economic opportunity and economic benefit into productive capacity into human competence into well-paid jobs, into wealth. What about non-tangible, non-economic benefits?"[41]

In sum, these ideas with a visualization potential culminated in 1991, i.e. suggestions about various expectations and responsibilities and both tangible and intangible outcomes matured and were consolidated by Carroll in his vision of the pyramid with four-layers of responsibilities which should overcome the apparent and/or alleged conflicts of a multi-stakeholder model.[42] Firstly, a business has the economic responsibility to be profitable, this is required by the society and this is clearly the *conditio sine qua non*—a business without any prospects of ever generating profits is not sustainable. Secondly, a business has a legal responsibility to obey the law, comply and fit in the legal framework and again this is required by the society—a business breaching the law on an ongoing basis does not belong in a state with the rule of law. Thirdly, a business has an ethical responsibility to distinguish between good and bad and to do what is just and fair and avoid harm. Unlike the

[38] Freeman (2017), pp. 449–465.

[39] Carroll (1991), pp. 39–48.

[40] Carroll (1979), pp. 497–505.

[41] Drucker (1984), pp. 53–63.

[42] Friske et al. (2020), pp. 138–155.

economic responsibility and legal responsibility, this is not, and cannot be, required by society. However, this is expected by society and often this expectation can reach the strength of economic and legal responsibilities, see e.g. how certain businesses are tabooed by their customers because they engage in some shady (but not strictly) illegal practices. Fourthly, a business has a philanthropic responsibility which is a mere wish generated by the society desiring that the business is a good corporate citizen and, as such, will show its generosity and concern for others.

This pyramid is a truly iconic depiction of social responsibility which is well recognized and updated on an ongoing basis, e.g. by recognizing the overlap of layers and the importance of the ethical layer. The study of all four layers, and in particular of two of them (ethical and philanthropic)[43] paved the venue to sharing values, aka CSV, which could be perceived as the, so far, highest stage of CSR. To put it differently, all four types of responsibilities have attracted attention, but systematically the most observed, discussed and reported are the economic and ethical responsibilities and it is assumed that they have a strong correlation with social demands, while less attention is paid to the legal and philanthropic responsibilities.[44]

The growing recognition and popularization of Carroll's pyramid allowed the UN to approach sustainable development as a conjunction of sustainability perceived as durability and of development perceived as extensiveness[45] and to turn it into a global command resting on three pillars: environmental (planet), social (people) and economic (profit), see Agenda 2030. Finally, it needs to be pointed out that Carroll's pyramid is not a static picture, it has been reviewed by its authors[46] as well as others. It is argued that ethics permeates the entire pyramid despite depicting the ethical responsibility as one single layer, because ethical considerations are present in each of the other responsibility categories as well.[47]

The transition of the CSR from a mere academic proposition to the sphere of business strategies and even state policies occurred in the EU in 2000 with the Lisbon Strategy,[48] and was cemented by the Green Paper for CSR in 2001.[49] Indeed, the Green Paper for CSR admitted that "20. Most definitions of corporate social responsibility describe it as a concept whereby companies integrate social and environmental concerns in their business operations and in their interactions with their stakeholders on a voluntary basis." In particular, pursuant to the Green Paper, CSR is "8. essentially a concept whereby companies decide voluntarily to contribute to a better society and a cleaner environment. At a time when the European Union

[43] Gössling and Vocht (2007), pp. 363–372.

[44] Gössling and Vocht (2007), pp. 363–372.

[45] Duran et al. (2015), pp. 806–811.

[46] Carroll and Shabana (2010), pp. 85–105.

[47] Carroll (2016), p. 3.

[48] Lisbon European Council on 23 and 24 March 2000 Presidency Conclusions ("Lisbon Strategy").

[49] COM(2001) 366 final Green Paper Promoting a European framework for Corporate Social Responsibility from 18 July 2001 ("Green Paper for CSR").

endeavours to identify its common values by adopting a Charter of Fundamental Rights, an increasing number of European companies recognise their social responsibility more and more clearly and consider it as part of their identity." Therefore, already in 2001, the CSR in the EU was not only strongly pro-value and pro-identity, but, as well, connected to the EU constitutional dimension, which is currently presented by the TEU,[50] TFEU[51] and Charter.[52]

Although CSR was voluntary in nature, public authorities in the EU started to take steps towards a pro-CSR policy and initiating legislation thereof.[53] Basically, in 2000 CSR was proclaimed, in 2001 linked to the constitutional dimension and in 2002 embedded in policies as precursors of the future EU legislation. Namely, already in 2002, the Commission presented its Communication about CSR.[54] Based on the collected feedback, the Communication about CSR confirmed that, despite the myriad of approaches to CSR, there is a consensus about three main features of CSR. "3. CSR is behaviour by businesses over and above legal requirements, voluntarily adopted because businesses deem it to be in their long-term interest; (ii) CSR is intrinsically linked to the concept of sustainable development: businesses need to integrate the economic, social and environmental impact in their operations; (iii) CSR is not an optional "add-on" to business core activities—but about the way in which businesses are managed." It is often overlooked, but the Communication about CSR, i.e. a policy document from 2002, already pushed the focus from large businesses to SMEs, see "4.5. The CSR concept was developed mainly by and for large multinational enterprises. In line with the Commission's "Think Small First" strategy, the CSR concept, practices and instruments should be adapted to suit the specific situation of SMEs which make up the vast majority of European enterprises. Because of their lower complexity and the strong role of the owner, SMEs often manage their societal impact in a more intuitive and informal way than large companies." The EU was very clear that CSR is neither a one size fits-all concept nor something only for big companies nor a strict frame of separated responsibilities. Hence, the development on the level of EU institutions fully matched the newest academic trends.

In 2006, along with the idea of the omnipresent and all penetrating ethical dimension, Michael Porter and Mark Kramer came up with the idea to overcome the eternal balancing and trade-offs and to move to a win-win model in the form of

[50]Treaty on EU (C 202/1, OJ C 202 7.6.2016, p. 13) – consolidated in 2020 ("TEU").

[51]Treaty on the Functioning of the EU (C 202/01, OJ C 202 7.6.2016, p. 47) – consolidated in 2020 ("TFEU").

[52]Charter of Fundamental Rights of the European Union (2016/C 202/02, OJ C 202, 7.6.2016, pp. 389–405) ("Charter").

[53]De Schutter (2008), pp. 203–236.

[54]COM(2002) 347 final Communication from the Commission concerning Corporate Social Responsibility: A business contribution to Sustainable Development from 2 July 2002 ("Communication about CSR").

connecting the competitive advantage[55] and corporate responsibility[56] leading to the CSV and diagrams further advancing the CSR pyramid. Based on the original version of these diagrams, CSV is the intersection of creating social value (addressing social and environmental objectives), typically CSR, and creating business value (being competitive and potentially profit generating). Based on a more advanced version of these diagrams, CSV is the intersection of business opportunities, social needs and corporate assets and expertise. Conventionally, three mechanisms are proposed for CSV: (i) reconceiving products and markets, (ii) redefining productivity in the value chain and (iii) enabling local cluster development. Plainly, CSR can and should contribute to a sustainable competitive advantage[57] and CSV should be about looking for opportunities and conditions how to achieve it.[58]

Similar to CSR, CSV is also a cost and, to make it a wise investment, it needs to be properly selected,[59] consistently applied and genuinely communicated.[60] As a matter of fact, the need for a proper choice of value and its non-hierarchical communications across the society cannot be over-emphasized.[61] Empirical studies point out that successful CSR, and, in particular, CSV initiatives, have one thing in common: they are aligned with the companies' business purpose.[62]

Arguably, a viable (if not the only one) option is to pragmatically and systematically tie both ends together, i.e. SDGs and social responsibility endeavours of businesses and individuals in an actionable manner and cross-sectorial partnership towards CSV,[63] otherwise the pro-sustainable investment is a mere waste.[64] This discussion about whether and what kind of CSR and CSV is effective and efficient became an integral part of conceptual considerations about the integration of CSR into the core business strategy.[65] Interestingly, this call for turning CSR/CSV from a debatable concept into a solid part of business conduct and business planning came from social scholars coming from different angles and sharing different views—Husted and Allen 2011,[66] Visser 2012,[67] etc.,—in the context of scepticism labelling traditional CSR to be a "confusing cliché which cheats consumers and society at

[55] Moon et al. (2014), pp. 205–225.
[56] Kramer and Pfitzer (2016), pp. 80–89.
[57] McWilliams and Siegel (2011), pp. 1480–1495.
[58] MacGregor Pelikánová et al. (2024b), pp. 1–29.
[59] Al-Jundi et al. (2019), pp. 1–2.
[60] MacGregor Pelikánová et al. (2024a), pp. 214–239.
[61] Ferraro and Beunza (2018), pp. 1187–1207.
[62] Rangan et al. (2015).
[63] Van Tulder and Van Mil (2023), pp. 1–6.
[64] Drucker (1973).
[65] Roszkowska-Menkes (2021).
[66] Husted and Allen (2011).
[67] Visser (2012).

large."[68] Later on, a term emerged for it, greenwashing, and exactly this is currently one of the key legislative targets of EU legislation.

8.3 CSR Stages, Phases and Versions

Considering the massive importance of CSR for sustainability and their mutual interaction, it would be remiss to end here and remain satisfied with the cursory summary that CSR means a pyramid of four categories of responsibilities, that CSV is its advanced form looking for a win-win solution and that somewhere, on the backstage, the shareholder theory and the stakeholder theory are fading away. CSR is undergoing a dynamic process, various lines are developed in parallel at various speeds and it is highly questionable whether the evolution is more linear or cyclical.

In the same way as the traditionalist shareholder theory and the stakeholder theory keep on evolving, so CSR keeps changing over time, in both its meaning and concepts. This rather complex evolution might be perceived as the motion between versions—from a very basic and rudimentary CSR to more advanced versions—and deserves a systemic presentation. Phases between these CSR versions can be understood as phases of societal and disciplinary thinking of CSR and can be instrumental in reaching a deeper understanding of CSR. Plainly, the CSR concept has been assigned different meanings over time for contextual, historical, disciplinary, political, economic and other reasons while the focus has gradually shifted towards a broader, developed, stricter and more far reaching version of responsibility.[69] In other words, from a single and narrow accountability to a multi-spectral liability and from a fiduciary duty focusing (only) on the shareholder value in the narrow sense to a fiduciary duty focusing on the value for the entire society. In sum, the current CSR appreciation demands the systemic study of the CSR 1.0 to CSR 4.0 categories.

CSR 1.0 (1953–1991) is labelled as Corporate Self Responsibility and basically represents the perceptions and suggestions from the initial launching decades of modern CSR, i.e. this is the CSR envisaged by Bowen, visualized by the early works of Carroll and directly confronted with Friedman's traditionalist shareholder theory. CSR 1.0 was the responsibility of businesses, as understood between 1953 and 1991, where the question was whether the fiduciary role of the management is to be judged from the perspective of only shareholders or, as well, stakeholders. Freeman's thought that there is a direct correlation between the financial performance and the relationship with the (primary) stakeholders was making its way. Contextually, it needs to be pointed out that the CSR 1.0 era occurred during the Third Industrial Revolution aka Industry 3.0 (1969–2000) which was marked by the use and development of nuclear energy and electronics. Finally, it needs to be emphasized that

[68] Ahen and Zettinig (2015), p. 95.

[69] Van Tulder and Van Mil (2023), pp. 327–336.

CSR 1.0 ended around the time of the fall of the iron curtain, the rise of globalization and the creation of the EU in Maastricht.

CSR 2.0 (1991–2006) is labelled as Corporate Social Responsiveness and represents the era when it was admitted that businesses are neither separated from society nor having an entirely independent life. Globalization made issues and crises omnipresent and businesses realized that they needed to build up their world-wide reputations, avoid any implication in scandals and boost their societal legitimacy. Doing something inappropriate could result in losing not only primary stakeholders, such as investors, employees and customers, but, as well, secondary stakeholders, such as suppliers, communities, etc. The importance of business ethics increased and it was acknowledged that CSR rests on two principles: the charity principle and the stewardship principle,[70] i.e. that privileged people should protect the interests of the less privileged and consider the entrusted confidence.[71] During this era, from 1991 to 2006, the Business in Society school provided theoretical foundations to this trend by supporting the connection between financial and social performance and by highlighting the critical role of corporate communications and business ethics.[72] This theory made its way into practice and management started to recognize the increasingly visible role of business in society[73] while following the social contract theory. The parallel with the sustainability could be observed while taking advantage of the iconic monograph, Cannibals with Forks, by Elkington from 1999,[74] which distinguishes three dimensions of CSR, aka a triple bottom line, aka People, Planet and Profit and was later on used as a source of the explanation of "new capitalism", even if this was not originally intended by Elkington.[75] The Business in Society perspective of CSR took on a life of its own and was about to move to a new level, CSR 3.0, which is dominated by the idea of CSV, in the context of the move from Industry 3.0 Electronic and Nuclear (1969–2000) to Industry 4.0 Internet and Renewable Energy (2000-).

CSR 3.0 (2006–2014) is labelled as Corporate Societal and Strategic Responsibility and represents the transition of CSR from a mere marketing-damage preventing tool to a business necessity, because of the link between CSR and the competitive advantage.[76] A business-society approach developed while demanding a systematic and structural interaction between businesses and stakeholders. Meanwhile, in reality, a closer scrutiny revealed that the level of trust in businesses to be effective, i.e. to do the right thing, decreased, especially damaging were the long lists of corporate scandals with international impacts. If CSR 1.0 and CSR 2.0 were strongly marked by the common law setting, then CSR 3.0 became much more

[70] Frederick et al. (1992).

[71] Van Tulder and Van Mil (2023), pp. 344–356.

[72] Kaptein and Wempe (2002).

[73] McWilliams and Siegel (2001), pp. 117–127.

[74] Elkington (1999).

[75] Van Tulder and Van Mil (2023), pp. 344–356.

[76] Porter and Kramer (2006).

influenced by the continental law attitude and by the Fourth Industrial Revolution—Industry 4.0 which brought about the shift to digitalization and energy renewability. Indeed, the era of CSR 3.0, i.e. the era from 2006 to 2014, was marked by the increasing activities of EU institutions, especially the European Commission. This is not surprising, especially considering a set of crises in 2007–2009, 2011, etc. The European Commission was behind a number of pro-CSR endeavors which culminated in a myriad of policies as well as legislative instruments. COM(2001) 366 final Green Paper Promoting a European framework for Corporate Social Responsibility provided the initial suggestions about CSR by stating that "Corporate social responsibility is essentially a concept whereby companies decide voluntarily to contribute to a better society and a cleaner environment." Five years later, COM(2006) 136 final Communication Implementing the Partnership for Growth And Jobs: Making Europe a Pole of Excellence on Corporate Social Responsibility, brought out the first EU definition of CSR by stating "Corporate social responsibility (CSR) is a concept whereby companies integrate social and environmental concerns in their business operations and in their interaction with their stakeholders on a voluntary basis." The next policy milestone was Europe 2020,[77] promoting, via its flagship initiatives, the renewal of the EU strategy to promote CSR.

These policy endeavours led to the critical change of the legislative panorama which, at least regarding reporting by certain subjects, started to turn the CSR into a liability. Namely, the Accounting Directive[78] was amended after one year by the NFDR,[79] which added Art. 19a about the Non-financial statement provided "1. Large undertakings which are public-interest entities exceeding on their balance sheet dates the criterion of the average number of 500 employees during the financial year shall include in the management report a non-financial statement containing information to the extent necessary for an understanding of the undertaking's development, performance, position and impact of its activity, relating to, as a minimum, environmental, social and employee matters, respect for human rights, anti-corruption and bribery matters, ...". Hence, the last year of CSR 3.0 was the very first year of a legal duty to report about CSR. Certainly, this legal duty was rather ambiguously stated, was imposed only unto a rather small pool of businesses and its enforcement was missing.[80] Nevertheless, it was a start and the cradle of the CSR focus categorization, i.e. the emergence of six categories of CSR concentration:

---

[77] COM(2010) 2020 Final Communication from the Commission: EUROPE 2020 A strategy for smart, sustainable and inclusive growth from 3 March 2010 ("Europe 2020").

[78] Directive 2013/34/EU of the European Parliament and of the Council of 26 June 2013 on the annual financial statements, consolidated financial statements and related reports of certain types of undertakings, amending Directive 2006/43/EC of the European Parliament and of the Council and repealing Council Directives 78/660/EEC and 83/349/EEC ("Accounting Directive").

[79] Directive 2014/95/EU of the European Parliament and of the Council of 22 October 2014 amending Directive 2013/34/EU as regards disclosure of non-financial and diversity information by certain large undertakings and groups ("NFRD").

[80] MacGregor Pelikánová and MacGregor (2020), pp. 74–108.

economic, environmental, employees, human rights, anti-corruption and R&D.[81] Since then, the EU law has kept growing in this arena, progressively turning not only the reporting duty but the entire CSR into a generally applicable legal duty for the financial sector,[82] see the SFDR,[83] or in general (for all sectors),[84] see the Taxonomy Regulation.[85] At the same time, the "old" Accounting Directive keep extending its reach and scope,[86] see the CSRD.[87] These new legislative trends reflect the development towards CSR 4.0.

CSR 4.0 (2014 -) is labelled as Corporate Societal and Sustainable Responsibility and represents the idea of mainstreaming sustainable business with a strong emphasis on the inclusion of societal goals and on contributing to a common purpose.[88] At the centre is the Agenda 2030 with 17 SDGs and the Paris Climate Agreement, both UN documents prepared with active support from the EU. CSR is not only about considering shareholders, primary stakeholders and secondary stakeholders, but rather it is also about considering the entire current society and even future stakeholders. The business ethics should be included in management processes,[89] the CSR is to be understood as an opportunity rather than an obligation or even a threat,[90] and a strategic, long-term, systematic approach is to be taken.[91] Ultimately, businesses are expected to do both—maintain profitability and behave responsibly in the larger sense.[92]

Consequently, CSR should be a bridge and constitute an intra-generational support, similar to the sustainability as envisaged by the SDGs. The battle should be for trust, i.e. the trust in creating and sharing genuine systemic values with an active support of tripartite collective action (social contract). The equation is simple—businesses cannot thrive in a society that fails and societies cannot be resilient where businesses blunder.[93] The multistakeholder model is the model for CSR. The age of greed with the defensive CSR 1.0, the age of marketing with

---

[81] MacGregor Pelikánová et al. (2021b), p. 3715.

[82] Rubáček et al. (2023), pp. 123–143.

[83] Regulation (EU) 2019/2088 of the European Parliament and of the Council of 27 November 2019 on sustainability-related disclosures in the financial services sector ("SFDR").

[84] MacGregor Pelikánová and Rubáček (2022), pp. 173–195.

[85] Regulation (EU) 2020/852 of the European Parliament and of the Council of 18 June 2020 on the establishment of a framework to facilitate sustainable investment, and amending Regulation (EU) 2019/2088 ("Taxonomy Regulation").

[86] MacGregor Pelikánová et al. (2024b), pp. 1–29.

[87] Directive (EU) 2022/2464 of the European Parliament and of the Council of 14 December 2022 amending Regulation (EU) No 537/2014, Directive 2004/109/EC, Directive 2006/43/EC and Directive 2013/34/EU, as regards corporate sustainability reporting ("CSRD").

[88] Van Tulder and Van Mil (2023), pp. 351–356.

[89] Sroka and Szántó (2018), pp. 111–126.

[90] Bhattacharya et al. (2011).

[91] Sroka and Lőrinczy (2015), pp. 156–163.

[92] Mohr et al. (2001), pp. 45–72.

[93] Van Tulder and Van Mil (2023), pp. 540–543.

promotional CSR 2.0 and the age of philanthropy with the charitable CSR 3.0 should be replaced by the age of responsibility, with systematic multi stakeholder-ism with interactive, strategic and transformative CSR 4.0. in the context of Industry 4.0, which should contribute to pro-sustainability uniform supply chain processes.[94]

8.4 Status quo of CSR

The Fourth Industrial Revolution, which is also referred to as Industry 4.0, leads to the current Society 4.0 which is defined as a group of people in advanced economies who form a modern cultural society, which is heavily influenced by technological developments and digitalization and having a significant impact on most aspects of human life.[95] The Society 4.0 should gradually transform in the Society 5.0 along with CSR. The near future will tell us how.

The present is showing that the conventional perception of the ultimate responsibility of a business (to increase the profits, preferably of shareholders)[96] has been questioned and there now prevails the opinion that CSR, which is neither a revolutionary idea nor a buzzword, is a meaningful, incremental scientific development, i.e. "CSR is the strategic process through which corporations can solve a social problem which is aligned to their value chain while pursuing economic profits."[97] In addition, a growing number of experts argue that CSR has a direct positive impact on the innovation, reputation[98] and even the business performance *per se*.[99] These proponents of CSR often assume that the implementation of CSR by businesses and their agents, entrusted to attain the mission of the business, results in the improvement of economic, environmental, social and governance dimensions, which in turn increases the satisfaction of all stakeholders.[100] At the same time, many studies point out that the positive relationship between CSR and financial performance is at least not obvious and even suggest that the improvement of CSR does not lead to an increase of total revenues of the business[101] and that the ethical rating of businesses often does not have any significant impact on the accounting or market ratio, except the negative impact in the case of a scandal.[102]

[94] Pickard et al. (2019), pp. 30–36.

[95] Turečková et al. (2023), pp. 146–157.

[96] Friedman (2007), pp. 173–178.

[97] Menghwar and Daood (2021), pp. 466–485.

[98] Kantabutra and Ketprapakorn (2020), pp. 1–22.

[99] Gallardo-Vázquez et al. (2019), p. 5614.

[100] Kantabutra and Ketprapakorn (2020), pp. 1–22.

[101] Matuszewska-Pierzynka (2021), pp. 885–906.

[102] Soana (2011), pp. 133–148.

Indeed, even the most enthusiastic proponents of CSR recognize that there are conditions and requirements to be met,[103] i.e. CSR must be selected and applied in a smart manner to be effective, efficient and legitimate[104] and must be properly communicated, see the dramatically important advertising intensity.[105] In addition, the pathway from the invention or other creative idea resolving a sustainability or some other kind of problem over its implementation via innovation to the ultimate progress, competitive advantage and embellished reputation is very complicated, complex and risky without any guarantee of success.[106]

The current literature unanimously demands the addressing of these apparent and/or alleged conflicts via a multi-stakeholder approach and/or model. The solution was presented as the four responsibilities layer pyramid of CSR[107] and, more recently, via various advanced CSV models[108] expecting a systematic and pragmatic approach.[109] Further, the literature is very clear that, regarding CSR, there is a need for a consistent endorsement[110] and a non-hierarchical communication across the society[111] and that a viable (if not the only one) option is to tie both ends together in a pro-active manner, i.e. SDGs and social responsibility endeavours of businesses and individuals in an actionable manner and cross-sectorial partnership.[112]

At the same time, despite this strong academic tenor boosted by massive policy and legislative endeavours orchestrated by the EU, the position of CSR is fragile. COVID-19 and post-COVID-19 events appear not to lead to boosting CSR and CSR support[113] by various stakeholders.[114] Similarly, the war in Ukraine, including the energy crisis and application of sanction packages,[115] appears to challenge, especially, certain CSR categories rather than support them, see e.g., the ongoing attack on the European Green Deal. In addition, generally well reported CSR (responsibility) has a positive impact on the firm value while generally well reported CSI (irresponsibility) has a negative impact on the firm value,[116] but it is possible that key stakeholders such as customers and investors will react in an opposed manner due to their social identification and other factors.[117] For example, the Volkswagen

---

[103] Kathayat (2022), pp. 701–714.
[104] MacGregor Pelikánová et al. (2024b), pp. 1–29.
[105] Matuszewska-Pierzynka (2021), pp. 885–906.
[106] MacGregor Pelikánová (2019), pp. 13–33.
[107] Friske et al. (2020), pp. 138–155.
[108] Kramer and Pfitzer (2016), pp. 80–89.
[109] Rangan et al. (2015).
[110] Tang et al. (2012), pp. 1274–1303.
[111] Ferraro and Beunza (2018), pp. 1187–1207.
[112] Van Tulder and Van Mil (2023).
[113] Hála et al. (2022), pp. 85–110.
[114] Hála et al. (2023), pp. 1–26.
[115] Malý et al. (2023), pp. 119–130.
[116] Ma and Xue (2023), pp. 3717–3738.
[117] Destri et al. (2024), pp. 1–26.

Dieselgate scandal led to a drop in the firms value and sales (customers and investors decided to punish) while the Parmalat scandal led to an increase in sales (customers and investors decided to support) and represents a "solidarity case."[118]

Many voices point out that the time is not ripe and that it is inefficient to exaggerate the effectiveness in the intra-related universe where many jurisdictions ignore, if not directly reject, the entire concept of sustainability with SDGs and CSR with CSV. Even the very bastion of the EU and its exclusive conferred competencies, the EU competition law, perceives sustainability and CSR rather as a shield than a sword[119] and to a certain extent delegates the CSR enforcement task to investors[120] and consumers.[121] Similarly, the EU law regarding intellectual property is hesitant regarding concerns for or against CSR.[122] The diverse readiness and maturity towards CSR can be observed across law fields as well as industries. For example, higher expectations are more likely to be expressed formally (by EU law) and informally (by various stakeholders) vis-à-vis large businesses than regarding SMEs,[123] and obviously many SMEs are merely hoping that CSRD will not hit them in the near future. Similarly, more so than other industries, the fashion industry appears to be directly targeted by both EU policies and law, such as policies fighting against waste and for recycling,[124] and to be closely scrutinized by stakeholders expecting to see some CSR action,[125] while other industries hope that "it will pass", see a segment from the hotel and catering industries.[126] The EU is pushing for the end of trade-offs and for genuine CSR to be communicated in a standardized form[127] via various channels to all.[128] However, the EU even pushes for various projects supporting sustainability and CSR, such as electric cars, which are not easily and pragmatically feasible and which do not have an undisputed way towards sustainability. Information-eager and discussion-ready investors and consumers,[129] in particular, advance their view of CSR, which is often different from the view of the EU as heralded by the internal EU pro-integration tandem, the European Commission and CJ EU.[130] This leads to a challenging and potentially controversial issue of European values, their determination and legal strength. It is argued that values such as human dignity, solidarity and social justice, environmental

[118]Destri et al. (2024), pp. 1–26.

[119]MacGregor Pelikánová (2024a), pp. 61–72.

[120]Rubáček et al. (2023), pp. 123–143.

[121]MacGregor Pelikánová (2024b), pp. 218–235.

[122]MacGregor Pelikánová and Beneš (2023), pp. 39–60.

[123]MacGregor Pelikánová and Zavřelová (2023), pp. 196–206.

[124]Cvik and MacGregor Pelikánová (2021), pp. 109–129.

[125]MacGregor Pelikánová and Sani (2023), pp. 813–851.

[126]MacGregor Pelikánová et al. (2021a), pp. 525–546.

[127]Balcerzak et al. (2023), pp. 199–212.

[128]MacGregor Pelikánová (2021), pp. 211–222.

[129]MacGregor Pelikánová and Hála (2021), p. 390.

[130]MacGregor Pelikánová and Hála (2021), p. 390.

sustainability, transparency and co-determination have not only direct impact on (at least some) private subjects of EU law, but even have an external perspective.[131] Well, the move of the discourse based on Art. 3 TEU and refereeing to solidarity and human right to the global dimension ultimately leads to the confrontation of dramatically different perspectives and perceptions regarding not only solidarity and human right, but also sustainability and CSR. The criticism after the Brundtland Report and arguments mentioning eco-imperialism, social-imperialism and human rights imperialism are very present and definitely not about to evaporate.

In sum, as before, CSR is multifaceted, continuously changing, and has various dimensions and several categories and its versions and visions are welcome by some and rejected by others. These dimensions and categories often lead to prioritizing and balancing and so, even fictively assuming the high maturity of the business regarding the CSR, this can lead to a fragmentation. The legal framework pushes certain aspects over others while each group of stakeholders has its own priorities, not to mention the interests of future stakeholders (which, of course, in CSR 4.0 must be considered). This inevitably leads to inconsistencies across domains and stakeholders.[132] The all-bridging synergetic idea of CSV as the top form of CSR meeting the expectations of all stakeholders is more of an illusion, perhaps even a chimera, than a reality. The growing integration of sustainability into business operations, as well as in EU policy and law frameworks, is probably of one of the major trends in CSR. CSR is expected to become an essential component of how businesses function as opposed to being considered as a distinct or optional activity.[133] However, each business has its own particularities and so each business has to adopt its own comprehensive and long-term view of its operations, and its impact, while taking into account current society as well as the future society. Western civilization is based on Christianity and its core values are creativity, private property and responsibility, and exactly this should determine the appreciation of CSR by not only businesses, but the entire society, which should go for an active and open-minded multi-stakeholder model.

Legal Documents

Code of Hammurabi, 1755 - 1750 B.C., a collection of 282 rules

Lex Duodecim Tabularum – Duodecim Tabulae (Law of the Twelve Tables), 449 BC

Corpus Iuris Civilis (Codex, Digest or Pandects, Institutes/Institutiones | Novellae Constitutiones), 529–534

[131] Frischhut (2019), pp. 35–36.

[132] Aguinis and Glavas (2012), pp. 932–968.

[133] Kharabsheh et al. (2023), p. 64.

An Acte againste suche persones as doo make Bankrupte 34 & 35, 8th Parliament of Henry VIII, c. 4 of 1542 ("Statute of Bankrupts of 1542")

Dampfkessel-Gesetzgebung in Preußen von 1831

Preußische Gewerbeordnung von 1845

Reichsgewerbeordnung von 1871

Gesetz, betreffend die Krankenversicherung der Arbeiter (KVG) vom 15.Juni 1883 - Deutsches Reichsgesetzblatt Band 1883, Nr. 9, Seite 73–104 ("German Health Insurance Act 1883")

Unfallversicherungsgesetz vom 6.Juli 1884 - Deutsches Reichsgesetzblatt Band 1884, Nr. 19, Seite 69 ("German Accident Insurance Act 1884")

Einschränkung der Frauenarbeit und Kinderarbeit Gesetz von 17.Juni 1887 Deutsches Reichsgesetzblatt Band 1887, Nr. 88, Seite 913 ("Working Hours Limitation Act 1887")

Gesetz betreffend die Invaliditäts- und Altersversicherung vom 22.Juni 1889 - Deutsches Reichsgesetzblatt Band 1889, Nr. 13, Seite 97 ("German Old Age and Invalidity Insurance Act 1889")

UN Annex to document A/42/427 Report of the World Commission on Environment and Development: Our Common Future from 4 August 1987 ("Brundtland Report")

Lisbon European Council on 23 and 24 March 2000 Presidency Conclusions ("Lisbon Strategy")

COM(2001) 366 final Green Paper Promoting a European framework for Corporate Social Responsibility from 18 July 2001 ("Green Paper for CSR")

COM(2002) 347 final Communication from the Commission concerning Corporate Social Responsibility: A business contribution to Sustainable Development from 2 July 2002 ("Communication about CSR")

COM(2010) 2020 Final Communication from the Commission: EUROPE 2020 A strategy for smart, sustainable and inclusive growth from 3 March 2010 ("Europe 2020")

Directive 2013/34/EU of the European Parliament and of the Council of 26 June 2013 on the annual financial statements, consolidated financial statements and related reports of certain types of undertakings, amending Directive 2006/43/EC of the European Parliament and of the Council and repealing Council Directives 78/660/EEC and 83/349/EEC ("Accounting Directive")

Directive 2014/95/EU of the European Parliament and of the Council of 22 October 2014 amending Directive 2013/34/EU as regards disclosure of non-financial and diversity information by certain large undertakings and groups ("NFRD")

Treaty on EU (C 202/1, OJ C 202 7.6.2016, p. 13) – consolidated in 2020 ("TEU")

Treaty on the Functioning of the EU (C 202/01, OJ C 202 7.6.2016, p. 47) – consolidated in 2020 ("TFEU")

Charter of Fundamental Rights of the European Union (2016/C 202/02, OJ C 202, 7.6.2016, pp. 389–405) ("Charter")

Regulation (EU) 2019/2088 of the European Parliament and of the Council of 27 November 2019 on sustainability-related disclosures in the financial services sector ("SFDR")

Regulation (EU) 2020/852 of the European Parliament and of the Council of 18 June 2020 on the establishment of a framework to facilitate sustainable investment, and amending Regulation (EU) 2019/2088 ("Taxonomy Regulation")

Directive (EU) 2022/2464 of the European Parliament and of the Council of 14 December 2022 amending Regulation (EU) No 537/2014, Directive 2004/109/EC, Directive 2006/43/EC and Directive 2013/34/EU, as regards corporate sustainability reporting ("CSRD")

OECD (2023), OECD Guidelines for Multinational Enterprises on Responsible Business Conduct, 8 June, 2023, OECD Publishing, Paris. https://doi.org/10.1787/81f92357-en ("OECD Guidelines 2023")

References

Aguinis H, Glavas A (2012) What we know and don't know about corporate social responsibility. J Manag 38:932–968

Ahen F, Zettinig P (2015) Critical perspectives on strategic CSR: what is sustainable value co-creation orientation? Crit Perspect Int Bus 11(1):92–109. https://doi.org/10.1108/MBE-09-2016-0047

Al-Jundi SA, Shuhaiber A, Augustine R (2019) Effect of consumer innovativeness on new product purchase intentions through learning process and perceived value. Cogent Bus Manag 6(1):1–2. https://doi.org/10.1080/23311975.2019.1698849

Balcerzak A, MacGregor RK, MacGregor Pelikánová R, Rogalska E, Szostek D (2023) The EU regulation of sustainable investment: the end of sustainability trade-offs? Entrep Bus Econ Rev 11(1):199–212. https://doi.org/10.15678/EBER.2023.110111

Bhattacharya CB, Sen S, Korschun D (2011) Leveraging corporate responsibility. The stakeholder route to maximizing business and social value. Cambridge University Press, Cambridge

Blake W 1804 (1970) Milton, a Poem. A. H. Bullen, London

Bowen HR (2013) Social responsibilities of the businessman. University of Iowa Press

Carroll AB (1979) A three-dimensional conceptual model of corporate performance. Acad Manag Rev 4(4):497–505. https://doi.org/10.2307/257850

Carroll AB (1991) The pyramid of corporate social responsibility: toward the moral management of organizational stakeholders. Bus Horiz 34(4):39–48

Carroll AB (2016) Carroll's pyramid of CSR: taking another look. Int J Corp Soc Responsib 1:3. https://doi.org/10.1186/s40991-016-0004-6

Carroll AB, Shabana KM (2010) The business case for corporate social responsibility: a review of concepts, research and practice. Int J Manag Rev 12(1):85–105. https://doi.org/10.1111/j.1468-2370.2009.00275.x

Carson SG, Øivind H, Sethi SP (2015) From implicit to explicit CSR in a Scandinavian context: the cases of HÅG and Hydro. J Bus Ethics 127(1):17–31. https://doi.org/10.1007/s10551-013-1791-2

Cvik ED, MacGregor Pelikánová R (2021) The significance of CSR during COVID-19 pandemic in the luxury fashion industry – a front-line case study. Eur J Bus Sci Technol 7(1):109–129. https://doi.org/10.11118/ejobsat.2021.005

D'Adamo I, Lupi G (2021) Sustainability and resilience after COVID-19: a circular premium in the fashion industry. Sustainability 13(4):1861. https://doi.org/10.3390/su13041861

De Schutter O (2008) Corporate social responsibility European style. Eur Law J 14(2):203–236. https://doi.org/10.1111/j.1468-0386.2007.00411.x

Destri MLD, Minà A, Picone PM (2024) Corporate social irresponsibility and stakeholders' support: evidence from a case study. J Manag Gov 28(2):1–26. https://doi.org/10.1007/s10997-022-09639-4
Drucker PF (1973) Management: tasks, responsibilities, practices. Harper & Row, New York
Drucker PF (1984) Converting social problems into business opportunities: the new meaning of corporate social responsibility. Calif Manag Rev 26(2):53–63
Duran DC, Gogan LM, Artene AE, Duran V (2015) The components of sustainable development - a possible approach. Procedia Econ Finance 26:806–811. https://doi.org/10.1016/S2212-5671(15)00849-7
Elkington J (1999) Cannibals with forks, the triple bottom line of the 21st century business. Capstone Publishing, Oxford
Ferraro F, Beunza D (2018) Creating common ground: a communicative action model of dialogue in shareholder engagement. Organ Sci 29(6):1187–1207. https://doi.org/10.1287/orsc.2018.1226
Frederick WC, Post JE, Davis K (1992) Business and society: corporate strategy, public policy, ethics. McGraw-Hill, New York
Freeman RE (1984) Strategic management – a stake- holder approach. Pitman, Boston
Freeman RE (2017) The new story of business: towards a more responsible capitalism. Bus Soc Rev 122(3):449–465. https://doi.org/10.1111/basr.12123
Friedman F (2007) The social responsibility of business is to increase its profits. In: Zimmerli WC, Richter K, Holzinger M (eds) Corporate ethics and corporate governance. Springer, Berlin, Heidelberg, pp 173–178
Friedrich O (1996) Blood & Iron. From Bismarck to Hilter. The Von Moltke Family's Impact on German History. HarperPerennial, New York
Frischhut M (2019) The ethical spirit of EU law. Springer, Cham
Friske W, Nikolov AN, Hoang P (2020) CSR reporting practices: an integrative model and analysis. J Market Theory Pract 28(2):138–155. https://doi.org/10.1080/10696679.2019.1690397
Gallardo-Vázquez D, Valder D, Valdez L, Castuera-Diáz AM (2019) Corporate social responsibility as an antecedent of innovation, reputation, and competitiveness success: a multiple mediation analysis. Sustainability 11(20):5614. https://doi.org/10.3390/su11205614
Gantt H (1919) Organizing for work. Harcourt, Brace and Howe, New York
Gössling T, Vocht C (2007) Social role conceptions and CSR policy success. J Bus Ethics 74:363–372. https://doi.org/10.1007/s10551-007-9512-3
Hála M, Cvik ED, MacGregor Pelikánová RM (2022) Logistic regression of Czech luxury fashion purchasing habits during the Covid-19 pandemic – old for loyalty and young for sustainability? Folia Oeconomica Stetinensia 22(1):85–110. https://doi.org/10.2478/foli-2022-0005
Hála M, MacGregor Pelikánová R, Rubáček F (2023) Negative determinants of CSR support by Generation Z in Central Europe - Infodemic's gender-sensitive impacts in a 'COVID-19' era. Central Eur Bus Rev 13(2):1–26. https://doi.org/10.18267/j.cebr.344
Hilger S (1998) Welfare policy in German big business after the first world. Bus Hist 40(1):50–76
Husted BW (2015) Corporate social responsibility practice from 1800-1914: past initiatives and current debates. Bus Ethics Q 25(1):125–141. http://www.jstor.org/stable/43695042
Husted BW, Allen DB (2011) Corporate social strategy. Cambridge University Press, Cambridge
Kantabutra S, Ketprapakorn N (2020) Toward a theory of corporate sustainability: a theoretical integration and exploration. J Clean Prod 270:1–22. https://doi.org/10.1016/j.jclepro.2020.122292
Kaptein M, Wempe J (2002) The balanced company. A theory of corporate integrity. Oxford University Press, New York
Kathayat A (2022) A case against CSR initiatives. Open J Bus Manag 10(2):701–714. https://doi.org/10.4236/ojbm.2022.102039
Kharabsheh B, Al-Shammari H, Bataineh K (2023) Research on corporate social responsibility: insights and future directions. Adm Sci 13(2):64. https://doi.org/10.3390/admsci13020064
Kramer MR, Pfitzer MW (2016) The ecosystem of shared value. Harv Bus Rev 94:80–89

Lauesen LM (2013) OECD principles of corporate governance and CSR. In: Idowu SO, Capaldi N, Zu L, Gupta AD (eds) Encyclopedia of corporate social responsibility. Springer, Heidelberg, pp 1771–1777. https://doi.org/10.1007/978-3-642-28036-8_25

Ma Y, Xue L (2023) Corporate social irresponsibility: a contingent social media crisis management perspective. Manag Decis 61(12):3717–3738. https://doi.org/10.1108/MD-07-2022-0986

MacGregor Pelikánová R (2019) R&D expenditure and innovation in the EU and selected member states. JEMI – J Entrep Manag Innov 15(1):13–33. https://doi.org/10.7341/20191511

MacGregor Pelikánová R (2021) Internal website presentation of Czech luxury fashion businesses in the Covid-19 era. Market Manag Innov 3:211–222. https://doi.org/10.21272/mmi.2021.3-18

MacGregor Pelikánová R (2024a) Chapter 5. Sustainability and competition law in Czech Republic. In: Këllezi et al (eds) Sustainability objectives in competition and intellectual property law, LIDC contributions on antitrust law, intellectual property and unfair competition. Springer, pp 61–72. https://doi.org/10.1007/978-3-031-44869-0_561

MacGregor Pelikánová R (2024b) EU legislative proposals to involve customers in the greening of the economy. Lawyer Q 2:218–235. https://tlq.ilaw.cas.cz/index.php/tlq/article/view/597

MacGregor Pelikánová R, Beneš M (2023) A turbulent pathway to uniform patent protection in the EU. Roma Tre Law Rev 2:39–60

MacGregor Pelikánová R, Hála M (2021) CSR unconscious consumption by Generation Z in the COVID-19 era – responsible heretics not paying CSR bonus? J Risk Financ Manag 14(8):390. https://doi.org/10.3390/jrfm14080390

MacGregor Pelikánová R, MacGregor RK (2020) The EU puzzling CSR regime and the confused perception by ambassadors of luxury fashion businesses: a case study from Pařížská. Central Eur Bus Rev 9(3):74–108. https://doi.org/10.18267/j.cebr.240

MacGregor Pelikánová R, Rubáček F (2022) Taxonomy for transparency in non-financial statements – clear duty with unclear sanction. Danube 13(3):173–195. https://doi.org/10.2478/danb-2022-0011

MacGregor Pelikánová R, Sani M (2023) Luxury, slow and fast fashion – a case study on the (un)sustainable creating of shared values. Equilibrium. Q J Econ Econ Policy 18(3):813–851. https://doi.org/10.24136/eq.2023.026

MacGregor Pelikánová R, Zavřelová V (2023) Values endorsed by top responsible large Czech companies – existent and inconsistent? In: Proceedings of the 16th International Conference: Liberec Economic Forum, Liberec, September 5–6, 2023, pp 196–206. https://doi.org/0.15240/tul/009/lef-2023-22

MacGregor Pelikánová R, Cvik ED, MacGregor RK (2021a) Addressing the COVID-19 challenges by SMEs in the hotel industry – a Czech sustainability message for emerging economies. J Entrep Emerg Econ 3(4):525–546. https://doi.org/10.1108/JEEE-07-2020-0245

MacGregor Pelikánová R, Němečková T, MacGregor RK (2021b) CSR statements in international and Czech luxury fashion industry at the onset and during the COVID-19 pandemic—slowing down the fast fashion business? Sustainability 13(7):3715. https://doi.org/10.3390/su13073715

MacGregor Pelikánová R, Nesbitt T, Balcerzak AP, Oulehla J (2024a) (In)effective communication about social responsibility? Examining large European businesses in the Czech Republic. Bus Manag Econ Eng 22(2):214–239. https://doi.org/10.3846/bmee.2024.21414

MacGregor Pelikánová R, Sani M, Rubáček F (2024b) Sustainable and responsible creation of shared values in the fast fashion industry – I will believe it when I see it. Vis Sustain 21(8873): 1–29. https://doi.org/10.13135/2384-8677/8873

Mähönen J, Johnsen G (2019) Law, culture and sustainability: corporate governance in the Nordic countries. In: Sjåfjell B, Bruner CM (eds) The Cambridge handbook of corporate law, corporate governance and sustainability. Cambridge law handbooks. Cambridge University Press, pp 218–231. https://doi.org/10.1017/9781108658386.022

Malý M, Cvik ED, MacGregor Pelikánová R (2023) EU sanctions against the Russian Federation and their implications for the foreign trade of the Czech Republic. AGRIS On-line Pap Econ Inform 15(3):119–130. https://doi.org/10.7160/aol.2023.150310

Matuszewska-Pierzynka A (2021) Relationship between corporate sustainability performance and corporate financial performance: evidence from U.S. companies. Equilibrium. Q J Econ Econ Policy 16(4):885–906. https://doi.org/10.24136/eq.2021.033

McCreary EC (1968) Social welfare and business: the Krupp Welfare Program, 1860-1940. Bus Hist Rev 42(1):24–49

McWilliams A, Siegel DS (2001) Corporate social responsibility: a theory of the firm perspective. Acad Manag Rev 26(1):117–127

McWilliams A, Siegel DS (2011) Creating and capturing value: strategic corporate social responsibility, resource-based theory, and sustainable competitive advantage. J Manag 37(5): 1480–1495

Menghwar PS, Daood A (2021) Creating shared value: a systematic review, synthesis and integrative perspective. Int J Manag Rev 23(4):466–485. https://doi.org/10.1111/ijmr.12252

Mohr LA, Webb DJ, Harris KE (2001) Do consumers expect companies to be socially responsible? The impact of corporate social responsibility on buying behavior. J Consum Aff 32(1):45–72. https://doi.org/10.1111/j.1745-6606.2001.tb00102.x

Moon HC et al (2014) Extending Porter's generic strategies: from three to eight. Eur J Int Manag 8(2):205–225. https://doi.org/10.1504/EJIM.2014.059583

Pickard M, Grecu I, Grecu G (2019) Sustainable smart manufacturing in Industry 4.0: real-time resource planning, process monitoring, and production control. Econ Manag Financ Markets 14: 30–36. https://doi.org/10.22381/EMFM14320194

Porter ME, Ignatius A (n.d.) Creating shared value: an HBR interview with Michael Porter (1/2). Available from http://www.youtube.com/watch?v=F44G4B2uVh4. Accessed 24 Feb 2024

Porter ME, Kramer MR (2006) Strategy and society: the link between competitive advantage and corporate social responsibility. Harvard Business Review, December

Rangan KV, Chase L, Sohel K. (2015) The truth about CSR – most of these programs are not strategic – and this is OK. Harvard Business Review, January – February, 2015

Roszkowska-Menkes M (2021) Strategic management and CSR. In: Idowu S et al (eds) Encyclopedia of sustainable management. Springer, Cham. https://doi.org/10.1007/978-3-030-02006-4_387-1

Rubáček F, MacGregor Pelikánová R, MacGregor RK (2023) The sustainability of ESAs triumvirate for sustainability-related disclosures in the financial sector – all for one and one for all. Acta Universitatis Lodziensis. Folia Iuridica 105:123–143. https://doi.org/10.18778/0208-6069.105.08

Soana M-G (2011) The relationship between corporate social performance and corporate financial performance in the bank sector. J Bus Ethics 104(1):133–148. https://doi.org/10.1007/s10551-011-0894-x

Spencer EG (1979) Rulers of the Ruhr: leadership and authority in German big businesses before 1914. Bus Hist Rev 53(1):40–64

Sroka W, Lőrinczy M (2015) The perception of ethics in business: analysis of research results. Procedia – Econ Finance 34:156–163. https://doi.org/10.1016/S2212-5671(15)01614-7

Sroka W, Szántó R (2018) Corporate social responsibility and business ethics in controversial sectors: analysis of research results. J Entrep Manag Innov (JEMI) 14(3):111–126. https://doi.org/10.7341/20181435

Stark A (1993) What's the matter with business ethics? Harv Bus Rev 71(3):38–48

Tang Z, Hull CE, Rothenberg S (2012) How corporate social responsibility engagement strategy moderates the CSR-financial performance relationship. J Manag Stud 49(7):1274–1303. https://doi.org/10.1111/j.1467-6486.2012.01068.x

Turečková K, Nevima J, Vaňová A, Vitálišová K (2023) Society 4.0: general economic implications. J Eur Econ 22(2):146–157. https://doi.org/10.35774/jee2023.02

Uekoetter F (1999) Divergent responses to identical problems: businessmen and the smoke nuisance in Germany and the United States, 1880-1917. Bus Hist Rev 73(4):641–676

Van Tulder R, Van Mil E (2023) Principles of sustainable business. Frameworks for corporate action on the SDGs. Routledge, London and New York

Visser W (2012) The age of responsibility: CSR 2.0 and the new DNA of business. Wiley, London

Chapter 9
Enforcement of Sustainability by Investors: Environment Social Governance

The modern history of sustainability is strongly shaped by the UN endeavours with a strong endorsement by the EU and is marked by the shift to the multi-sectorial globalization[1] requiring the employment of the multi-stakeholder approach.[2] Plainly, the involvement of private law subjects is indispensable and the trend of the dominance from the shareholder theory (1950–1984) over to the stakeholder theory (1984–1991), with trade-offs to multi-responsibilities and win-win, is obvious. This is the scenery of CSR moving to CSV in order to make sure that each and every member of society is involved and contributes to sustainability, resting on three pillars and linked to SDGs. However, such an endeavour entails both rights and duties and the resulting benefits do not need to be immediate or tangible. Plainly, going for sustainability while engaging with CSR can, at least *prima facie*, create costs to be carried not only by the business itself but as well by its stakeholders and ultimately should bring benefits to all of them, and even the entire society. Such a mechanism is, in Western civilization, feasible only based on the free will of the subjects and consequently Sustainability European Style depends upon proper information inducing the subjects to be sustainable and go for sustainability. As a matter of fact, the only two other options, ordering sustainability by mandatory law provisions or even managing centrally sustainability by the state, are hardly to be reconciled with European values and do not fully meet the test of the rule of law and the economic rule of the free market. So, for better or for worse, Sustainability European Style needs ESG to be endorsed by businesses and shared by other informed stakeholders. Indeed, these stakeholders, especially the almighty investors, could become the most powerful, effective and efficient enforcers of sustainability at the private law level. Plainly, a business violating environmental and social demands and behaving contrary to CSR might risk to be cut off, not only by disappointed clients and no longer trusting suppliers, but as well by finance providers. The destiny

[1] Schüz (2012), pp. 7–15.

[2] Van Tulder and Van Mil (2023), pp. 379–393.

R. MacGregor Pelikánová, *Sustainability in Europe*,
https://doi.org/10.1007/978-3-031-74689-5_9

of such a business could be sealed and ultimately the free hand of the market with its index finger labelled ESG, could erase it from the business scene and sink it into oblivion.

ESG means considering environmental issues, social issues and corporate governance while making and carrying investments. Such an investment is to be responsible and with a pro-sustainability impact. Investment decisions have always focused on the likelihood (risk) and rate of return in the financial sense, but almost never on that alone. Indeed, the reasons for investment have always been heterogeneous and often political. Other consideration could even overshadow a simple return calculation. However, a systematic linking and tying of sustainability projected into CSR to investment has a history of only half a century. Firstly, the idea that, while operating a business, not just shareholders should be considered but that attention should be paid to investors as powerful stakeholders as well, needed to be established and cemented during 1972–2004 (Sect. 9.1). Then, it was possible to engage in the discussion about how, i.e. whether the market self-regulation could lead to the very needed ESG during 2004–2015 (Sect. 9.2). Finally, similarly to CSR, the last stage came in 2015 and means the UN standardization (Sect. 9.3), which developed particularly under the auspices of the EU law (Sect. 9.4).

9.1 Rediscovery of the Investor as a Stakeholder—CSR Not Only for Shareholders—ESG (1972–2004)

The origins of ESG are linked to the social responsibility investing movement in the 1960s and 1970s[3] and to one highly influential book. Indeed, in 1972, Meadows published The limits to growth, which expressed clearly that tangible resources are neither unlimited nor self-recharging, that they are becoming exhausted and that they must be balanced with the increasing world population.[4] If one uses something (tangible), then there might be not enough left for the other one. Around this time, the era of postwar Keynesianism endorsing government intervention and generally supporting the idea of a strong and big state was getting replaced by the neoliberalist economic theory, preferring privatization, deregulation, austerity and a reduction in government spending.[5]

The Austrian anti-Keynesian economist and polymath, Friedrich A. Hayek, had vindicated individual liberty in all spheres of public life.[6] Consequently, he demanded even smaller government actions (*laisser faire* with a mere observance of the rule of law) reflecting the Austrian school.[7] He was blamed for engaging in an

[3]Zatonatska et al. (2024), p. 2950.

[4]Meadows et al. (1972).

[5]Balcerzak and MacGregor Pelikánová (2020), p. 95.

[6]Lehmann (2014), pp. 104–117.

[7]MacGregor Pelikánová and Hála (2021), p. 390.

utopic contradiction of a free society pursuing (complicated and complex) mandates and at the same time consisting of humans with insufficient and/or fragmented knowledge. Ironically, this utopia has been, allegedly, fully realized via powerful neoliberal regimes, such as Reagan's USA and Thatcher's UK, relying on privatization and financial market deregulation.[8] Manifestly, key market players, businesses and their investors, needed to be at least partially committed and informed.

The genuine popularization of neoliberal thought was launched by Milton Friedman from the University in Chicago, who did not hesitate to position himself at the nexus of an influential new group of transatlantic conservative thinkers generating policy agendas for the Reagan and Thatcher revolutions. It was Milton Friedman who transformed and converted Adam Smith's classical doctrine of the invisible hand, referring to self-interested benign and social pay-off actions, into a direct attack on the public sphere, see his quote in the context of the two-hundredth anniversary of the Smith's *magnum opus,* An Inquiry into the Nature and Causes of the Wealth of Nations, i.e. that "the invisible hand in politics, that is the precise reverse of the invisible hand in the market."[9] Since the destiny of the society, including economics, should not be so much in the hands of a government imposing strict regulations, somebody else should do the job and have the responsibility.

Milton Friedman and his wife, Rose Friedman, with their best-selling book Free to Choose: A Personal Statement, from 1980, argued for a reduction of the role and involvement of government, reflecting the Chicago school and contributed to the creation of the shareholder theory demanding that businesses do legally all possible to maximize the profits of shareholders, later on slightly moderated towards the motto "what is good for business is good for society."[10] While Hayek distrusted the allure of popularity and feared oversimplification, the robustly entrepreneurial Friedman embraced it and became the key figure of the era of CSR 1.0. In other words, Friedman was a leading proponent of the shareholder theory, with the shareholder wealth maximization imperative motivated by the agency problem and hazards arising from the separation of risk bearing and decision-making (also known as ownership v control).[11]

Sustainability was projected into CSR 1.0 (1953–1991) which was labelled as Corporate Self Responsibility and which judged the fiduciary duty of the management either from the perspective of shareholders alone, see Friedman's shareholder theory, or stakeholders, see Freeman's stakeholder theory. In the context of discussions about the business performance and fiduciary duty, there came the proposition that there is a correlation between the financial performance and the Corporate Self Responsibility, i.e. even during CSR 1.0 it was argued that socially and environmentally responsible behaviour might lead to economic profits. Consequently, the

[8] Lehmann (2014), pp. 104–117.

[9] Lehmann (2014), pp. 104–117.

[10] Porter and Ignatius (n.d.).

[11] Phillips et al. (2003), pp. 479–502.

information about such a responsibility might be relevant for an investor interested in the economic profits of the business.

In 1991, the Rubicon was crossed and the increasingly globalized society moved to CSR 2.0 (1991–2006), which is labelled as Corporate Social Responsiveness and which recognized that business is an integral part of the society, i.e. the business depends upon the existence of the society and so destroying the society means the self-destruction of the business. The recognition of the mutual existence interdependence boosted the importance of the moral dimension and reputation and made it clear that the societal legitimacy of the business excludes anti-social and anti-environmental behaviour. This led to the stakeholder theory, which combines the organizational management and ethics and consequently deals with moral topics and leads to moral or immoral solutions, but is never amoral.[12] Plainly, the stakeholder theory addresses explicitly both morals and value.[13] The key figure of the stakeholder theory is R. Edward Freeman, who worked extensively on its moral foundation and who referred both to the deontological Kantianism and to the concept of fairness.[14] In this context, CSR is a form of business self-regulation which forms an integral part of the business model. The fiduciary duty is due not just to shareholders, and the goal is more than a mere maximizing of shareholder wealth.[15] This departure and/or dilution of the focus on shareholder profits *stricto sensu* naturally brings the danger of managerial opportunism. Ultimately, the stakeholder theory is close to alternative models of strategic management, such as the resource dependence theory,[16] and leads to a better balancing of decisions than the shareholder theory.

If, during CSR 1.0, investors might have considered non-financial information about the business in order to calculate the risk and amount of their return, then, during CSR 2.0, investors had to have considered it, not just might. Unethical behaviour and violations of the charity principle and the stewardship principle were, pursuant to the Business in Society school, very strong, if not conclusive, reasons to not invest in such a business, because the financial and social performances are ultimately connected.[17] In sum, investors are to be considered.

In 1999, John Elkington published his iconic monograph, Cannibals with Forks,[18] which distinguishes three dimensions of CSR, aka a triple bottom line—People, Planet and Profit, and argued that social justice, economic prosperity and environmental quality are key criteria to assess the business (corporate)

[12] Freeman (1994), pp. 409–442.

[13] Phillips et al. (2003), pp. 479–502.

[14] Freeman (1994), pp. 409–442.

[15] Phillips et al. (2003), pp. 479–502.

[16] Frooman (1999), pp. 191–205.

[17] Painter-Morland (2013), pp. 283–293.

[18] Elkington (1998).

performance.[19] He accepted and endorsed the idea that corporations can lead society towards a sustainable future, but has not specified how.[20]

Further, towards the end of CSR 2.0 it was recognized that investors play a vital role and deserve legal protection, because the well-enforced rights of investors increase their willingness to finance firms and support the development of the financial market.[21] So, not only is the CSR information important for the investors assessing the risk and return of their investment, but, additionally, it is one of the pre-requirements for the development of the financial market. Plainly, CSR not only needs to be done, but needs to be reported as being done. Indeed, the Friedman's shareholder theory is for a defensive sustainability and CSR, because they protect shareholder value, which depends upon the investment potential. Consequently, the investors needed to know about a (lack of) potential environment hazards and, later on, with CSR 2.0 and in the context of the Brundtland Report from 1987, they needed to know about all the risks generated by the business with an impact on the sustainability of the entire society, i.e. not only environmental but, as well, social.[22] In addition, they needed such all of this information in a "measurable" and "quantifiable" form. Consequently, ESG criteria evolved significantly due to the transition from a niche interest into a central consideration in global investment and corporate strategies.[23] Non-financial factors presented via ESG criteria became necessary to identify material risks and growth opportunities as well as to predict financial performance.[24] Such a connection of non-financial factors and financial performance implied the demand for solid indexes and indices, and so in 1999 the Dow Jones Sustainability Indexes and London FTSE4-Good were established.[25]

9.2 CSR to Be Done and Said—Search for the ESG Regime (2004–2015)

Due to the frequent criticism that CSR is merely used as a public relations tool, if not directly a green washing device which does not genuinely address non financial issues, it became clear that a neutral and legitimacy-boosting mechanism needed to be used.[26] Firstly, CSR 3.0, dominated by the idea of CSV, replaced CSR 2.0, and secondly a metrics-based approach emerged to increase the corporate

[19] Rowlands (1999), pp. 42–43.

[20] Rowlands (1999), pp. 42–43.

[21] La Porta et al. (2002), pp. 1147–1170.

[22] Keeble (1988), pp. 17–25.

[23] Zatonatska et al. (2024), p. 2950.

[24] Friede et al. (2015), pp. 210–233.

[25] Gössling and Vocht (2007), pp. 363–372.

[26] Myers and Czarnezki (2021), pp. 991–1040.

accountability—the ESG index and scoring.[27] The iconic intangible and intuitive statement "sustainability: I know it when I see it",[28] needed to be materialized and quantified in order to overcome the ambiguity and subjectivism added by the overlap of CSR and CSV.[29] In 1987, the Brundtland Report defined sustainable development as "development that meets the needs of the present generation without compromising the ability of future generations to meet their own needs" suggesting that the instantaneous balancing of current stakeholders is insufficient. Businesses should not only be accountable to all stakeholders, including investors, but in addition they should align their operations and decisions in accordance with standardized expectations connecting the past, the present and the future—ISO 26000.[30] In the following years, Carroll updated his four layer responsibilities pyramid[31] and Porter and Kramer brought out CSV.[32] This led to the strategic and systemic, aka carrot-sermons, CSR 3.0 reflecting both environmental and social aspects[33] and interweaving them with economic development.[34] This CSR 3.0 became massively endorsed by businesses[35] with various levels of genuineness. Thus was created a need for a regime entailing transparent and reliable information sharing, especially regarding financial markets and investment—standardized and communicated ESG.

This need and idea were not entirely new and, especially in the USA, a myriad of the pre-cursors of ESG policies have mushroomed since the end of the nineteenth century. Naturally, these investment policies reflected contextually the social and other preferences taboos of the given era. Consequently, investments and investment funds have considered factors and commodities such as weapons, alcohol, tobacco as well as the provenience, such as South Africa, etc. A relevant legislation, turning these policies into legally enforceable instruments with the general application, started to emerge approximately five decades ago, see the Community Reinvestment Act 1977".[36] This Act is a seminal piece of legislation intended to address inequities in access to credit and to overcome racial discrimination in real estate and lending, which kept going on despite the Fair Housing Act.[37] Previous civil rights laws focused on negative rights and recommendations, i.e. on what lenders should not do, but the Community Reinvestment Act 1977 focussed on positive rights and recommendations, i.e. on what lenders should do—invest in their communities even if they

[27] Aggarwal and Kadyan (2014), pp. 61–66.

[28] White (2013), pp. 213–217.

[29] MacGregor Pelikánová and Sani (2023), pp. 813–851.

[30] Aggarwal and Kadyan (2014), pp. 61–66.

[31] Carroll (1999), pp. 268–295.

[32] Porter and Kramer (2006).

[33] Van Der Heijden (1999), pp. 199–221.

[34] O'Riordan (1985), pp. 1431–1446.

[35] Barbier (1987), p. 101.

[36] Community Reinvestment Act, 12 U.S.C. § 2901 et seq. from 12th October 1977 ("Community Reinvestment Act 1977").

[37] Fair Housing Act, 42 U.S.C. 3601 et seq., from 11th April 1968 ("Fair Housing Act 1968").

are low-income areas. US federal banking agencies have implemented this legislation and have monitored, in their light, the lending practices of banks.

However, these legal rules about sustainable investments and ESG were applicable only in one given jurisdiction, typically one sharing the common law tradition. The internationalization in this respect was induced only in 1987 with the Brundtland Report, by which the UN clearly took the initiative and has kept it until now.

In 2004 occurred a milestone regarding the development of ESG and its "worldwide" standardized regime. Namely, ESG became known due to a report prepared by financial institutions and presented under the title The Global Compact: Who Cares Wins. Connecting Financial Markets to a Changing World, aka Report Who Cares Wins,[38] in New York in 2004. The preparation of the Report Who Cares Wins was invited by the UN, in particular the UN Secretary-General Kofi Annan who asked 55 CEOs to draft guidelines on how to better integrate environmental, social and corporate governance issues into companies and the investment sector.[39] Namely, 18 institutions from 9 jurisdictions accepted such an invitation and together worked on guidelines and recommendations on how to better integrate ESG issues in asset management, securities brokerage services and associated research functions. The project was financed by the Swiss Government and the motivation behind it was the desire to boost the effectiveness, efficiency and legitimacy of sustainability reporting and green marketing. The resulting Report Who Cares Wins contains guidelines and recommendations on ESG, while emphasizing the importance of ESG for investors and investment.[40] One of the key points was expressed in the Executive summary part of the Report Who Cars Wins, namely on pages ii and iii, i.e. that "Financial institutions should commit to integrating environmental, social and governance factors in a more systematic way in research and investment processes… Companies are asked to take a leadership role by implementing environmental, social and corporate governance principles and polices and to provide information and reports on related performance in a more consistent and standardised format. They should identify and communicate key challenges and value drivers and prioritise environmental, social and governance issues accordingly. We believe that this information is best conveyed to financial markets through normal investor relation communication channels and encourage, when relevant, an explicit mention in the annual report of companies. … Investors are urged to explicitly request and reward research that includes environmental, social and governance aspects and to reward well-managed companies."

[38] The Global Compact: Who Cares Wins. Connecting Financial Markets to a Changing World from 24 June 2004 ("Report Who Cares Wins") - Recommendations by the financial industry to better integrate environmental, social and governance issues in analysis, asset management and securities brokerage. Available at https://www.unepfi.org/fileadmin/events/2004/stocks/who_cares_wins_global_compact_2004.pdf.

[39] Pollman (2022), 659/2022.

[40] Hilson (2024).

The Report Who Cares Wins was prepared in order to establish a non-contentious system for addressing material investment risks[41] and at the same time being perceived as neutral, unlike CSR which was subjected by Friedman and other shareholder theory proponents to a criticism for (allegedly) politicizing corporate decision making.[42] It became a springboard for the issuance of "world-wide" principles for reporting about sustainability for investors, which started to be discussed and drafted in the aftermath of the Report Who Cares Wins.

In 2005, the UN Secretary-General, Kofi Annan, and the initiator of the Report Who Cares Wins, invited the largest institutional investors world-wide to discuss and identify key principles regarding sustainable investments. On 27th April 2006, the UN Secretary-General Kofi Annan announced, together with a group of the world's largest institutional investors, six resulting principles for responsible investment, which were endorsed at the New York Stock Exchange aka Six Principles.[43] These Six Principles brought a global and general framework for investors to consider ESG issues. The idea behind it was that environmental issues, such as climate change, and social and political issues, such as human rights and their violation, can influence the performance of investment portfolios and can have a strong bearing even on narrowly defined financial results. Since both financial and non-financial (ESG) factors need to be considered by investors, accountability regarding ESG factors is an integral part of the fiduciary role played by both, managers towards various stakeholders, and investors. The leitmotif of these Six Principles is to inspire and induce the inclusion of ESG in investment analysis and decision-making processes and the general implementation. Further, the UN supported the establishment of an international network of financial institutions endorsing and supporting the implementation of these Six Principles, called the Principles for Responsible Investment ("UN PRI"). These Six Principles are a mere recommendation and their endorsement is fully voluntary. Nevertheless, they are aspirational and often expected, especially since the pool of UN PRI participants keeps growing, in particular in the context and aftermath of various financial and other crises, such as in 2008–2009. UN PRI is moving towards institutionalization and has offices with a staff in various cities word-wide. In addition, UN PRI has launched various ESG and Six principles promoting initiatives, including the UN PRI Academy, providing ESG training, and keeps issuing various reports, such as reports about the fiduciary duty.

While UN PRI has been working on spreading and implementing the Six Principles, the Global Reporting Initiative ("GRI"), a non-governmental organization founded in 1997 in Boston, has been issuing sustainability standards ("GRI Standards") regarding climate change, human rights, and corruption. The GRI Standards are either universal or industry specific. They are developed and updated by the Global Sustainability Standards Board and have a modular structure. The GRI

---

[41] Hilson (2024).

[42] Blumberg (1971), pp. 425–462.

[43] Puaschunder (2016), pp. 38–63.

perceives sustainability reporting as a pro-active instrument supporting sustainable development. It is a method of measuring, disclosing, and being accountable to internal and external stakeholders for the performance while considering sustainable development. GRI Standards are dominated by the double-materiality concept in the context of sustainability reporting. Exactly this was the inspiration for the European Commission in 2019 regarding not only policy, but as well law proposals.

As a matter of fact, the period of 2006–2015 was not just the era of CSR 3.0, but as well the era of a very intensified search for a standardization of CSR measurements, in particular for investors (ESG) via numerous UN and private initiatives. For example, in the USA, the Federal Trade Commission started to issue environmental marketing guides (green guides) already in 1992, and their revised version from 2012 provided detailed guidance regarding the information entailing non-deceptive and authentic environmental claims, i.e. against greenwashing.[44] Indeed, the action of the Federal Trade Commission shows the bridging between policies and law, while each jurisdiction with its policies and law was determining its own ESG regime. In 2008, No. 1276 The Business Protection from Misleading Marketing Regulations, which prohibits misleading advertising, was launched in the UK, while the EU and Nordic countries advanced various certification systems, see the Nordic Ecolabel and the EU Ecolabel. International standardization was involved partially, see ISO 14001 which assesses effective business environmental management. Manifestly, within the common law tradition, the roots of the national law regimes of ESG are in tort, while, in the continental law tradition, in the law against unfair competition and intellectual property law.

The beneficiaries of this standardization should be investors and such a transparent, reliable and compatible ESG (reporting) should improve the operation of financial markets, see the single internal EU market with the famous four freedoms of movement, and ultimately should bring benefits to all stakeholders, perhaps even future stakeholders. However, the era of 2006–2015 brought a set of criticisms of the emerging ESG and ESG reporting. The selected criteria and factors were questioned, their measurement methodology contested, the legitimacy was challenged and even the conceptual understanding of CSR and social good was discussed. Often at the centre of this criticism was the correct argument that, although investment decisions are chiefly based on the potential for financial (monetarily measurable) return, they have always reflected a myriad of other, heterogenous, considerations, such as political influence, ethical redemption, strategic acquisition, an increase of the intellectual property leverage, personal preferences, etc.

Well, regardless of these objections and the fragmentation of pertinent law regimes and standardization, the trend was clear and then became even further magnified in the context of crises with a strong financial dimension during 2007–2009 as well as of crises linked to the climate change. Around this time, effective risk management started to focus on traditional financial risks but as well

[44] Aggarwal and Kadyan (2014), pp. 61–66.

on ESG.[45] Therefore, ESG and ESG practices came not only as the sword to fight for CSR,[46] but as well as the shield, i.e. as preventive measures against possible damages caused in operation, marketing and even litigation area.[47] Ideally, the ESG as both a sword and shield should be holistically and systematically incorporated in the business strategy[48] and create a value.

By 2015, it was clear that ESG is relevant not only for the financial industry and that, despite the lack of universal standards, reported high ESG standards indicated a good CSR, an attractive stakeholder-friendly approach and a lower risk portfolio, as much as it concerns environmental and social issues. The financial industry across the world came to the same conclusion—regardless of the presence and absence of a pertinent law, that ESG data must be employed into investment decisions. The businesses needing to obtain investments got this message, as well as their partners, suppliers, consumers, etc.

By 2015, the majority of large businesses in the world were reporting about their sustainability performance, CSR strategy and generally how they impact the society at large. One common measure to understand, study and compare these reports became the ESG index.[49] However, at the same time, many malpractices and paradoxes emerge, such as, even though investors realized that ESG investments should yield higher returns, in reality the cost of having a fully integrated ESG strategy can offset the financial benefits deriving from those investments.[50] DESG (Dominantly ESG) related investing represents a complex task studied under severe information shortages.[51] Although some businesses have moved to greenwashing or cherry-picking or outsourcing their CSR through third parties[52] or merely passing on real evidence,[53] the CSR performance, described via ESG, the ESG index and ESG scoring, became a strong consideration. Consequently, by 2015, the global vox populi demanded even more ESG than ever before and wanted directly a systemic standardization of ESG.

[45] Teng et al. (2023), pp. 511–549.

[46] Qoyum et al. (2022), pp. 306–320.

[47] Reber et al. (2022) ESG, pp. 867–886.

[48] Harjoto and Laksmana (2018), pp. 353–373.

[49] Cayón and Gutierrez (2021), pp. 114–123.

[50] Cappucci (2018), pp. 22–28.

[51] Škapa et al. (2022), pp. 763–775.

[52] Cayón and Gutierrez (2021), pp. 114–123.

[53] Cappucci (2018), pp. 22–28.

9.3 The UN in Action—A Systemic Standardization of ESG by 2015 and After

The year 2015 represents a true milestone for sustainability, CSR and even ESG. The UN Agenda 2030 with SDGs came and changed both the public law and private law landscape. It cemented the perception of the rather organic and ongoing evolution of sustainability.[54] Further, it contributed to the move towards CSR 4.0, the recognition that CSV is a progressive stage of CSR which builds organically upon pre-existing thoughts.[55] In sum, by 2015, there were no longer any doubts that CSR had become an essential part of competitive strategy[56] and, at the same time, that CSR has many meanings, variations and shades which should be appreciated in the context of each particular industry and jurisdiction. Plainly, CSR endeavours entail many activities which bring different costs and a diverse likelihood of results. Plainly, the same type and category of CSR can lead to an increase in competitiveness, brand recognition and various stakeholder's satisfaction in one industry and not in another, in one jurisdiction and not another. Numerous recent studies have addressed it, e.g., regarding banking,[57] and the fashion industry.[58] Similarly, numerous recent studies have shown that pro-sustainable and pro-CSR behaviour might come up against the issue of the size, see SMEs,[59] or various industry, market, firm-specific and other barriers.[60] Hence, despite the positive rhetoric, sustainability and SDGs are not going through an entirely successful era since 2015 and even leading personalities, including those from the UN, admit it.[61] Many owners and managers are reluctant and very cautious about investing in CSR due to the disadvantages and costs[62] it may impose upon the structuring of their resources and operational processes.[63] Therefore, on one hand there is a demand for circularity and a proliferation of green finance instruments in order to improve the social setting and smooth over the transition towards a carbon negative world,[64] but on the other hand a pragmatic re-calculation is needed and can lead to support of its attitude[65] as well as to a more cautious approach.[66] Plainly, balancing economic, societal and environmental

[54] Kuhn (1970), p. 92.

[55] Porter and Kramer (2014), pp. 149–151.

[56] Nicole et al. (2022), pp. 1–14.

[57] Ivascu et al. (2023), pp. 1–18.

[58] MacGregor Pelikánová et al. (2024b), pp. 1–29.

[59] Ernst et al. (2022), pp. 1–14.

[60] Pham et al. (2022), pp. 2714–2735.

[61] MacGregor Pelikánová et al. (2024b), pp. 1–29.

[62] Ernst et al. (2022), pp. 1–14.

[63] Barauskaite and Streimikiene (2020), pp. 278–287.

[64] Aneja et al. (2023), pp. 42214–42222.

[65] Carrascosa-Lopez et al. (2012), pp. 567–574.

[66] D'Adamo et al. (2023), pp. 840–852.

objectives is a challenge and struggle for each and every business in a particular and specific manner.[67] This challenge and struggle should be approached in a systematic manner and ideally based on an economic model moving from linearity to circularity[68] with the aim of resource efficiency, long-term value creation, waste minimizing, reduction of closed loops and products considering environmental protection.[69]

This makes the enhancement of awareness about the heterogeneity of businesses and CSR indispensable, as well as the providing of the information about it with an incentive effect. Manifestly, businesses need to receive an additional push, and this push can be given by investors, provided the investors can learn about it and appreciate it. This, is already suggested by the Report Who Cares Win from 2004, but this works only if the expectations of investors and businesses meet and there is not an overly dramatic asymmetry of information. Plainly, standardized data, not only about financial indicators but as well about non-financial indicators, are an integral part of the investment decision-making process. This is fully compatible with the vision that CSV is not so much a revolutionary concept,[70] but instead an organic result of prior theories and models such as the pyramid or CSR,[71] which builds organically upon pre-existing thoughts[72] and leads to advanced CSV models.[73]

Obviously, there is a need for a genuine and non-hierarchical communication[74] across the society[75] and that a viable (if not the only one) option is to tie together both ends, i.e. SDGs and social responsibility endeavors of businesses and individuals, in an actionable manner and cross-sectorial partnership.[76] The conventional perception of the ultimate responsibility of a business (to increase the profits, preferably of shareholders)[77] has been questioned and now there prevails the opinion that CSR, which is neither a revolutionary idea nor a buzzword, is a meaningful, incremental scientific development, i.e. CSR is the strategic process through which corporations can solve a social problem, which is aligned to their value chain, while pursuing economic profits.[78]

As already mentioned, the development of sustainability/ESG reporting standards is predominantly linked to the GRI, which was developed in 1997 by the US non-profits coalition Ceres and the consulting agency Tellus Institute. GRI and its

---

[67] Eikelenboom and de Jong (2019), pp. 1360–1370.
[68] Skvarciany et al. (2021), pp. 11–34.
[69] Morseletto (2020), p. 104553.
[70] Wójcik (2016), pp. 32–55.
[71] Strand et al. (2015), pp. 1–15.
[72] Porter and Kramer (2014), pp. 149–151.
[73] Kramer and Pfitzer (2016), pp. 80–89.
[74] MacGregor Pelikánová et al. (2024a), pp. 214–239.
[75] Ferraro and Beunza (2018), pp. 1187–1207.
[76] Van Tulder and Van Mil (2023), pp. 9–11.
[77] Friedman (2007), pp. 173–178.
[78] Menghwar and Daood (2021), pp. 466–485.

endeavours have been receiving an ongoing support by the UN, in particular from the United Nations Environment Programme. Consequently, as much as GRI is institutionalized and independent, GRI co-operates with the UN, especially with the UN Environment program and the UN Global Compact. UN Secretary-General Kofi Annan described it as forming a "unique contribution to make in fostering transparency and accountability of corporate activities beyond financial matters".[79] The GRI global multi-stakeholder network has been providing a platform for analysis and feedback on the Guidelines and has contributed to the expansion of the GRI scope from merely environmental reporting to the employment of three sets of sustainability indicators: social performance indicators, economic performance indicators and environmental performance and impacts indicators. In sum, the GRI and the Sustainability Accounting Standards Board ("SASB") demonstrates two key approaches to materiality, with differences that may cause confusion in interpreting information about "material sustainability issues".[80] ESG metrics by GRI and ISOs are complementary and mutually influenced by (see GRI v "environmental" ISOs), such as the following, ISO 14010, ISO 14011, ISO 14012 and ISO 26000 and with the influence of OECD guidelines. Nevertheless, despite the massive popularity, it must be underlined that the compliance with GRI Standards (as well as with Six Principles) is voluntary. Similarly, the IFRS Foundation is a not-for-profit, public interest organization established to develop globally accepted accounting and sustainability disclosure standards while using the International Accounting Standards Board ("IASB") and International Sustainability Standards Board ("ISSB"). However, neither the IFRS Foundation nor IASB nor ISSB have the authority to dictate and impose these standards to subjects against their will. As a matter of fact, even the Agenda 2030 is not per se an enforceable mandatory instrument imposing duties to private subjects and so SDG metrics and calculations are rather tools to encourage managers and other stakeholders to adopt social and green solutions.[81] At the same time, it needs to be observed that a significant part of stakeholders expects, sometimes even demands, the employment of techniques for measuring and assessing the SDG performance of the given business and takes in consideration its results,[82] e.g. along with a more typical ESG data and other more "exotic" indicators, such as e.g. the Equitable and Sustainable Wellbeing indicators, aka ESW indicators.[83]

Hence, there is an obvious space for misinformation and manipulation and reverse spillover effects,[84] see greenwashing, cherry-picking or even more advanced and premediated social and/or environmental cleaning,[85] supply chains

[79] Brown et al. (2009), pp. 182–200.

[80] Jørgensen et al. (2022), pp. 341–361.

[81] D'Adamo et al. (2024), p. 5049.

[82] Siddiqui et al. (2023), pp. 91761–91779.

[83] D'Adamo et al. (2024), p. 5049.

[84] Rao (2022), pp. 534–554.

[85] Cayón and Gutierrez (2021), pp. 114–123.

manipulation with predatory practices[86] and the extremely repugnant and totally trust destroying disinformation.[87] Well, only as long as the law does not step in .... In order to circumvent "greenwashing" or falsified reporting, the EU as well as other states crossed the Rubicon and moved on to a change of law. This change of law targeted the genuine and measurable information about the sustainability and CSR behaviour of not only the business itself stricto sensu, but as well the due diligence regarding its business or supply chains[88] and generally all information needed to assess responsible investing, impact investing and potential risks to investment from all spheres.

9.4 The EU in Action—The Law on ESG by 2015 and After

By 2015, the EU law constitutional trio clearly recognized the integral position of sustainability by stating that the EU "shall work for the sustainable development of Europe based on balanced economic growth and price stability, a highly competitive social market economy, aiming at full employment and social progress, and a high level of protection and improvement of the quality of the environment" (Art. 3(3) TEU). The Charter added that "a high level of environmental protection and the improvement of the quality of the environment must be integrated into the policies of the Union and ensured in accordance with the principle of sustainable development" (Art. 37 Charter).

The pathway from constitutional values and principles, referring to sustainable development, to the standardized and compulsory ESG reporting, started to be built around the same time as the UN moved, with notable help from the EU, to SDGs and to focus both on general as well as sectorial (reporting aspects). Indeed, in 2014–2015, a number of events occurred in the EU and they had a massive impact on CSR and ESG—the Commission of José Manuel Barroso was replaced by the Commission of Jean-Claude Juncker, the War in Crimea started, the EU outside as well as inside trade brought new trends,[89] etc. In such a potentially unstable time, the EU decided to move to legislative instruments to consolidate and ideally boost the sustainability and the responsibility of a growing circle of subjects for it ... and this trend continues until now.

Firstly, there is a line of Directives addressing financial and non-financial reporting, including CSR, launched by the NFRD[90] which added to the Accounting

[86] Stepien and Polcyn (2019), pp. 382–388.

[87] Bennett and Livingston (2018), pp. 122–139.

[88] Millard (2021), pp. 54–63.

[89] Taušer et al. (2015), pp. 17–37.

[90] Directive 2014/95/EU of the European Parliament and of the Council of 22 October 2014 amending Directive 2013/34/EU as regards disclosure of non-financial and diversity information by certain large undertakings and groups ("NFRD").

Directive[91] the famous Art. 19a Non-financial statement—"Large undertakings which are public-interest entities exceeding on their balance sheet dates the criterion of the average number of 500 employees during the financial year shall include in the management report a non-financial statement containing information to the extent necessary for an understanding of the undertaking's development, performance, position and impact of its activity, relating to, as a minimum, environmental, social and employee matters, respect for human rights, anti-corruption and bribery matters, ..." From its very beginning, the NFRD demanded large companies to provide non-financial disclosures to the markets and to report on environmental, social and employee-related, human rights, anti-corruption and bribery matters as well as about their business models, outcomes and risks of the policies on the above topics, and the diversity policy applied for management, etc.

Neither the Accounting Directive nor the NFRD address directly in their provisions the term "sustainable development" and even, unlike already by then emerging legislation in other states, such as the UK,[92] did not address issues in business and supply chains. However, in its Preamble, the NFRD mentions "sustainable growth" and "sustainable global economy by combining long-term profitability with social justice and environmental protection." A set of further amendments of the Accounting Directive has followed and the last of them is the CSRD,[93] which amends the Accounting Directive as regards corporate sustainability reporting. Indeed, the CSRD refers directly to the European Green Deal and expands the duty set by Art. 19a as well as the pool of its addressees—to even smaller businesses and their suppliers. The CSRD mentions "sustainable development" in its Preamble and in relation to SDGs, see namely, "Sustainability reporting standards should also take account of internationally recognised principles and frameworks on responsible business conduct, corporate social responsibility, and sustainable development, including the SDGs,..." (Preamble at 45. CSRD). However, there is one direct reference to the sustainable development in the very body of the CSRD, namely the CSRD adds to the Accounting Directive a new provision about Sustainability Reporting Standards, see " When adopting delegated acts pursuant to paragraph 1, the Commission shall, to the greatest extent possible, take account of: (a) the work of global standard-setting initiatives for sustainability reporting, and existing standards and frameworks for natural capital accounting and for greenhouse gas accounting, responsible business conduct, corporate social responsibility, and sustainable development;..." (Art. 29b (5) Accounting Directive). Indeed, the CSRD fights against non-financial reporting fragmentation, refers to the Green Deal and the

[91] Directive 2013/34/EU of the European Parliament and of the Council of 26 June 2013 on the annual financial statements, consolidated financial statements and related reports of certain types of undertakings, amending Directive 2006/43/EC of the European Parliament and of the Council and repealing Council Directives 78/660/EEC and 83/349/EEC ("Accounting Directive").

[92] Modern Slavery Act 2015 c.30 from 26 March 2015.

[93] Directive (EU) 2022/2464 of the European Parliament and of the Council of 14 December 2022 amending Regulation (EU) No 537/2014, Directive 2004/109/EC, Directive 2006/43/EC and Directive 2013/34/EU, as regards corporate sustainability reporting ("CSRD").

SFDR and subjects a growing number of European Companies to reporting according to European Sustainability Reporting Standards ("ESRS"), i.e. making Environmental, Social, Governance (ESG) reporting the widespread reality of the current EU. In addition, such a reporting has to entail not only the data directly related to the pertinent business and its business conduct *stricto sensu*, but as well about the value chain and supply chain, see the, by CSRD updated, Art. 19a(2)f) (ii) "a description of "the principal actual or potential adverse impacts connected with the undertaking's own operations and with its value chain, including its products and services, its business relationships and its supply chain, actions taken to identify and monitor those impacts..." Therefore, CSRD addresses CSR via ESG in a holistic and systematic manner and fights against misleading, static and only one element of the chain presenting strategies, which are popular in the agriculture and food sectors[94] as well as other sectors and industries.

These ESRS take account of discussions under the auspices of the IFRS Foundation, in particular the ISSB, and the above mentioned GRI and their wording is incorporated in the EU law via the ESRS Regulation,[95] see in particular Annex I of the ESRS Regulation. Regarding the composition and focus of the ESRS, two out of these 12 standards are general, five environmental, four social, and one economic-governance.

Secondly, there is a line of sectorial and terminological regulations critical for ESG and ESG reporting. The SFDR,[96] initiated by the European Commission of Jean-Claude Juncker, targets malpractices and misleading information in the financial sector.[97] The SFDR focuses only on financial market participants and financial advisers and deals with the transparency of the provided information about sustainability risks and adverse sustainability impacts. Therefore, sustainable development is overshadowed by a focus on CSR information, i.e. the data about the sustainable investment, which is defined as "an investment in an economic activity that contributes to an environmental objective, as measured, for example, by key resource efficiency indicators on the use of energy, renewable energy, raw materials, water and land, on the production of waste, and greenhouse gas emissions, or on its impact on biodiversity and the circular economy, or an investment in an economic activity that contributes to a social objective, in particular an investment that contributes to tackling inequality or that fosters social cohesion, social integration and labour relations, or an investment in human capital or economically or socially disadvantaged communities, provided that such investments do not significantly harm any of those objectives and that the investee companies follow good governance practices,

[94] Stepien and Polcyn (2019), pp. 382–388.

[95] Commission Delegated Regulation (EU) 2023/2772 of 31 July 2023 supplementing Directive 2013/34/EU of the European Parliament and of the Council as regards sustainability reporting standards ("ESRS Regulation").

[96] Regulation (EU) 2019/2088 of the European Parliament and of the Council of 27 November 2019 on sustainability-related disclosures in the financial services sector ("SFDR").

[97] Rubáček et al. (2023), pp. 123–143.

..." (Art. 2(17) SFDR). Consequently, due to the SFDR, the sustainability development trade-offs start to be expressly and explicitly rejected. This was further magnified by a pro-Green Deal Taxonomy Regulation,[98] initiated by the European Commission of Ursula von der Leyen,[99] Indeed, the Taxonomy Regulation fiercely tackles advanced sustainability misinformation and disinformation, in particular greenwashing.[100] The Taxonomy Regulation mentions specifically sustainable development in its preamble while referring to SDGs and to technical screening criteria for the transport sector which contributes close to 26% of the total greenhouse gas emissions in the EU. It defines environmentally sustainable economic activities as activities contributing substantially to one or more of six environmental objectives without significantly harming any of them (Art. 3 Taxonomy Regulation). These six environmental objectives include (a) climate change mitigation; (b) climate change adaptation; (c) the sustainable use and protection of water and marine resources; (d) the transition to a circular economy; (e) pollution prevention and control; (f) the protection and restoration of biodiversity and ecosystems (Art. 9 Taxonomy Regulation). Hence the principle "do not significantly harm" launched by the SFDR is becoming very clear both regarding both the active (to do) as well as the passive (to omit) dimensions. In sum, the message conveyed by the SFDR and the Taxonomy Regulation is that the sustainable development is to be multi-spectral and without trade-offs. In addition, as mentioned above, the CSRD forces more and more businesses to reflect it and to speak about it while considering themselves as well as their suppliers and partners, i.e. see the ESG reporting duty. The CSRD keeps extending its pool of subjects having to report, while the European Commission keeps adopting standards, i.e. the ESRS, and even issuing further legislative instruments, such as the ESRS Regulation[101] supplementing the Accounting Directive as regards sustainability reporting standards. Since 2015, the Sustainability European Style definitely counts on investors being provided with the mandatory information from businesses. This is not merely about responsibilities, this is also about liabilities and the EU has correctly engaged with the legislation to increase the quantity, quality and access to such information. However, the EU should think twice before going for a massive mandatory legislation in this arena and should rather appreciate and take advantage of the already existing fine mechanism of the ostracism and solidarity exercised by well informed customers and investors, see e.g. the situation of scandals and CSI, which could lead to a punishment of the business by the large public (Volkswagen Dieselgate scandal),[102] but which could as well lead to a second

[98] Regulation (EU) 2020/852 of the European Parliament and of the Council of 18 June 2020 on the establishment of a framework to facilitate sustainable investment, and amending Regulation (EU) 2019/2088 ("Taxonomy Regulation").

[99] MacGregor Pelikánová and Rubáček (2022), pp. 173–195.

[100] Balcerzak et al. (2023), pp. 199–212.

[101] Commission Delegated Regulation (EU) 2023/2772 of 31 July 2023 supplementing Directive 2013/34/EU of the European Parliament and of the Council as regards sustainability reporting standards ("ESRS Regulation").

[102] Ma and Xue (2023), pp. 3717–3738.

chance and a massive solidarity and/or support due to the social identification (Parmalat scandal).[103] After all, sustainability is about the mutual aspiration and support to do better and this should be rather stimulated than ordered by the legislation. A carrot can be both more effective and efficient than a hardly definable and problematically enforceable stick. You catch more flies with honey than with vinegar. Well, the EU has a newly announced strategic plan with an attractive and promising title—A new plan for Europe's sustainable prosperity and competitiveness, announced by the New Political Guidelines.[104] Consequently, we shall soon see whether the scenery is becoming more pro-sustainability, pro-SDG, pro-CSR and pro-ESG … or less. Sustainability, I know it[105] and believe it,[106] when I see it.

Legal Documents

Fair Housing Act, 42 U.S.C. 3601 et seq., from 11th April 1968 ("Fair Housing Act 1968")

Community Reinvestment Act, 12 U.S.C. § 2901 et seq. from 12th October 1977 ("Community Reinvestment Act 1977")

Global Compact: Who Cares Wins. Connecting Financial Markets to a Changing World. Recommendations by the financial industry to better integrate environmental, social and governance issues in analysis, asset management and securities brokerage from 24 June 2004 ("Report Who Cares Wins"). Available at https://www.unepfi.org/fileadmin/events/2004/stocks/who_cares_wins_global_compact_2004.pdf

Modern Slavery Act 2015 c.30 from 26 March 2015

Directive 2013/34/EU of the European Parliament and of the Council of 26 June 2013 on the annual financial statements, consolidated financial statements and related reports of certain types of undertakings, amending Directive 2006/43/EC of the European Parliament and of the Council and repealing Council Directives 78/660/EEC and 83/349/EEC ("Accounting Directive")

Directive 2014/95/EU of the European Parliament and of the Council of 22 October 2014 amending Directive 2013/34/EU as regards disclosure of non-financial and diversity information by certain large undertakings and groups ("NFRD")

Modern Slavery Act 2015 c.30 from 26 March 2015

[103] Destri et al. (2024), pp. 1–26.

[104] Von der Leyen U (2024) Political Guidelines for the next European Commission 2024–2029: Europe's Choice, 18 July 2024. Available from: https://commission.europa.eu/document/e6cd4328-673c-4e7a-8683-f63ffb2cf648_en ("New Political Guidelines").

[105] White (2013), pp. 213–217.

[106] MacGregor Pelikánová et al. (2024b), pp. 1–29.

Regulation (EU) 2019/2088 of the European Parliament and of the Council of 27 November 2019 on sustainability-related disclosures in the financial services sector ("SFDR")
Regulation (EU) 2020/852 of the European Parliament and of the Council of 18 June 2020 on the establishment of a framework to facilitate sustainable investment, and amending Regulation (EU) 2019/2088 ("Taxonomy Regulation")
Directive (EU) 2022/2464 of the European Parliament and of the Council of 14 December 2022 amending Regulation (EU) No 537/2014, Directive 2004/109/EC, Directive 2006/43/EC and Directive 2013/34/EU, as regards corporate sustainability reporting ("CSRD")
Commission Delegated Regulation (EU) 2023/2772 of 31 July 2023 supplementing Directive 2013/34/EU of the European Parliament and of the Council as regards sustainability reporting standards ("ESRS Regulation")
Von der Leyen U (2024) Political Guidelines for the next European Commission 2024–2029: Europe's Choice, 18 July 2024. Available from: https://commission.europa.eu/document/e6cd4328-673c-4e7a-8683-f63ffb2cf648_en ("New Political Guidelines")

References

Aggarwal P, Kadyan A (2014) Greenwashing: the darker side of CSR. Indian J Appl Res 4(3): 61–66. https://doi.org/10.15373/2249555X/MAR2014/20
Aneja R, Kappil SR, Das N et al (2023) Does the green finance initiatives transform the world into a green economy? A study of green bond issuing countries. Environ Sci Pollut Res 30:42214–42222. https://doi-org.infozdroje.czu.cz/10.1007/s11356-023-25317-w
Balcerzak A, MacGregor Pelikánová R (2020) Projection of SDGs in codes of ethics – case study about lost in translation? Adm Sci 10(4):95. https://doi.org/10.3390/admsci10040095
Balcerzak A, MacGregor RK, MacGregor Pelikánová R, Rogalska E, Szostek D (2023) The EU regulation of sustainable investment: the end of sustainability trade-offs? Entrep Bus Econ Rev 11(1):199–212. https://doi.org/10.15678/EBER.2023.110111
Barauskaite G, Streimikiene D (2020) Corporate social responsibility and financial performance of companies: the puzzle of concepts, definitions and assessment methods. Corp Soc Responsib Environ Manag 28(1):278–287. https://doi.org/10.1002/csr.2048
Barbier EB (1987) The concept of sustainable economic development. Environ Conserv 14:101. https://doi.org/10.1017/S0376892900011449
Bennett WL, Livingston S (2018) The disinformation order: disruptive communication and the decline of democratic institutions. Eur J Commun 33(2):122–139. https://doi.org/10.1177/0267323118760317
Blumberg P (1971) The politicalization of the corporation. Bus Law Rev 51:425–462
Brown HS, de Jong M, Lessidrenska T (2009) The rise of the Global Reporting Initiative: a case of institutional entrepreneurship. Environ Polit 18(2):182–200. https://doi.org/10.1080/09644010802682551
Cappucci M (2018) The ESG Integration Paradox. J Appl Corp Finance 30(2):22–28. https://doi.org/10.1111/jacf.12296
Carrascosa-Lopez C, Segarra-Ona MV, Peiro-Signes A, Segura-García-del-Río B (2012) Environmental proactivity in manufacturing industries: is it valuable? Int Bus Econ Res J (IBER) 11(5): 567–574. https://doi.org/10.19030/iber.v11i5.6974

Carroll AB (1999) Corporate social responsibility: evolution of a definitional construct. Bus Soc 38(3):268–295. https://doi.org/10.1177/000765039903800303

Cayón E, Gutierrez JC (2021) Sin stocks and ESG scores: does the nature of your business really matter? J Int Stud 14(3):114–123. https://doi.org/10.14254/2071-8330.2021/14-3/7

D'Adamo I, Gastaldi M, Ozturk I (2023) The sustainable development of mobility in the green transition: renewable energy, local industrial chain, and battery recycling. Sustain Dev 31(2): 840–852. https://doi.org/10.1002/sd.2424

D'Adamo I, Di Carlo C, Gastaldi M, Rossi EN, Uricchio AF (2024) Economic performance, environmental protection and social progress: a cluster analysis comparison towards sustainable development. Sustainability 16:5049. https://doi.org/10.3390/su16125049

Destri MLD, Minà A, Picone PM (2024) Corporate social irresponsibility and stakeholders' support: evidence from a case study. J Manag Gov 28(2):1–26. https://doi.org/10.1007/s10997-022-09639-4

Eikelenboom M, de Jong G (2019) The impact of dynamic capabilities on the sustainability performance of SMEs. J Clean Prod 235:1360–1370. https://doi.org/10.1016/j.jclepro.2019.07.013

Elkington J (1998) Cannibals with forks: the triple bottom line of 21st century business. New Society Publishers, Gabriola Island, BC

Ernst RA, Gerken M, Hack A, Hülsbeck M (2022) SMES' reluctance to embrace corporate sustainability: the effect of stakeholder pressure on self-determination and the role of social proximity. J Clean Prod 335:1–14. https://doi.org/10.1016/j.jclepro.2021.130273

Ferraro F, Beunza D (2018) Creating common ground: a communicative action model of dialogue in shareholder engagement. Organ Sci 29(6):1187–1207. https://doi.org/10.1287/orsc.2018.1226

Freeman RE (1994) The politics of stakeholder theory: some future directions. Bus Ethics Q 4(4): 409–442

Friede G, Busch T, Bassen A (2015) ESG and financial performance. Aggregated evidence from more than 2000 empirical studies. J Sustain Financ Invest 5:210–233. https://doi.org/10.1080/20430795.2015.1118917

Friedman M (2007) The social responsibility of business is to increase its profits. In: Zimmerli WC, Richter K, Holzinger M (eds) Corporate ethics and corporate governance. Springer, Berlin, Heidelberg, pp 173–178

Frooman J (1999) Stakeholder influence strategies. Acad Manag Rev 24(2):191–205

Gössling T, Vocht C (2007) Social role conceptions and CSR policy success. J Bus Ethics 74:363–372. https://doi.org/10.1007/s10551-007-9512-3

Harjoto M, Laksmana I (2018) The impact of corporate social responsibility on risk taking and firm value. J Bus Ethics 151(2):353–373. https://doi.org/10.1007/s10551-016-3202-y

Hilson Ch (2024) Climate change and politicization of ESG in the US. Front Polit Sci 6, 15 March 2024. https://doi.org/10.3389/fpos.2024.1332399

Ivascu L, Ali W, Khalid R, Raza M (2023) The impact of competitive strategies on. Performance of banking sector; the mediating role of corporate social responsibility and operational excellence. Energies 16(1):297. https://doi.org/10.3390/en16010297

Jørgensen S, Mjøs A, Pedersen LTJ (2022) Sustainability reporting and approaches to materiality: tensions and potential resolutions. Sustain Account Manag Policy J 13(2):341–361. https://doi.org/10.1108/SAMPJ-01-2021-0009

Keeble BR (1988) The Brundtland Report: 'Our Common Future'. Med War 4(1):17–25. https://www.jstor.org/stable/45353161

Kramer MR, Pfitzer MW (2016) The ecosystem of shared value. Harv Bus Rev 94:80–89

Kuhn TS (1970) The structure of scientific revolutions. University of Chicago Press, Chicago

La Porta R, Lopez-de-Silanes F, Shleifer A, Vishny R (2002) Investor protection and corporate valuation. J Finance 57(3):1147–1170. http://www.jstor.org/stable/2697775

Lehmann C (2014) Neoliberalism, the revolution in reverse. The Baffler 24:104–117. http://www.jstor.org/stable/43306894

Ma Y, Xue L (2023) Corporate social irresponsibility: a contingent social media crisis management perspective. Manag Decis 61(12):3717–3738. https://doi.org/10.1108/MD-07-2022-0986

MacGregor Pelikánová R, Hála M (2021) CSR unconscious consumption by Generation Z in the COVID-19 era – responsible heretics not paying CSR bonus? J Risk Financ Manag 14(8):390. https://doi.org/10.3390/jrfm14080390

MacGregor Pelikánová R, Rubáček F (2022) Taxonomy for transparency in non-financial statements – clear duty with unclear sanction. Danube 13(3):173–195. https://doi.org/10.2478/danb-2022-0011

MacGregor Pelikánová R, Sani M (2023) Luxury, slow and fast fashion – a case study on the (un)sustainable creating of shared values. Equilibrium. Q J Econ Econ Policy 18(3):813–851. https://doi.org/10.24136/eq.2023.026

MacGregor Pelikánová R, Nesbitt T, Balcerzak AP, Oulehla J (2024a) (In)effective communication about social responsibility? Examining large European businesses in the Czech Republic. Bus Manag Econ Eng 22(2):214–239. https://doi.org/10.3846/bmee.2024.21414

MacGregor Pelikánová R, Sani M, Rubáček F (2024b) Sustainable and responsible creation of shared values in the fast fashion industry – I will believe it when I see it. Vis Sustain 21(8873): 1–29. https://doi.org/10.13135/2384-8677/8873

Meadows DH, Meadows DL, Randers J, Behrens WW (1972) The limits to growth. Universe Books, New York

Menghwar PS, Daood A (2021) Creating shared value: a systematic review, synthesis and integrative perspective. Int J Manag Rev 23:466–485. https://doi.org/10.1111/ijmr.12252

Millard E (2021) How Fair Trade can stay strong as companies scale up their social and environmental programmes. J Fair Trade 3(1):54–63. https://doi.org/10.13169/jfairtrade.3.1.0054

Morseletto P (2020) Targets for a circular economy. Resour Conserv Recycl 153:104553. https://doi.org/10.1016/j.resconrec.2019.104553

Myers C, Czarnezki JJ (2021) Sustainable business law? The key role of corporate governance and finance. Environ Law 51(4):991–1040. https://www.jstor.org/stable/48647568

Nicole S, Lada S, Ansar R, Adis AA, Ming Fook L, Chekima B (2022) Corporate social responsibility and strategic management: a bibliometric analysis. Sustainability 14(17):1–14. https://doi.org/10.3390/su141710526

O'Riordan T (1985) Research policy and review 6. Future directions for environmental policy. Environ Plann A: Econ Space 17:1431–1446. https://doi.org/10.1068/a171431

Painter-Morland M (2013) The role of business in society. In: Osburg T, Schmidpeter R (eds) Social innovation. CSR, sustainability, ethics & governance. Springer, Berlin, Heidelberg, pp 283–293. https://doi.org/10.1007/978-3-642-36540-9_25

Pham H, Pham T, Dang CN (2022) Barriers to corporate social responsibility practices. Eng Const Arch Manag 29(7):2714–2735. https://doi.org/10.3390/buildings13061561

Phillips R, Freeman RE, Wicks AC (2003) What stakeholder theory is not. Bus Ethics Q 13(4): 479–502. http://www.jstor.org/stable/3857968

Pollman E (2022) The making and meaning of ESG. U. Of Penn., Inst. For law & Econ. Research paper no. 22-23, European corporate governance institute - law working paper no. 659/2022. Available at: https://ssrn.com/abstract=4219857

Porter ME, Ignatius A (n.d.) Creating shared value: an HBR interview with Michael Porter (1/2). Available from http://www.youtube.com/watch?v=F44G4B2uVh4. Accessed 24 Feb 2024

Porter ME, Kramer MR (2006) Strategy and society: the link between competitive advantage and corporate social responsibility. Harvard Business Review, December

Porter ME, Kramer MR (2014) A response to Andrew Crane et al.'s article. Calif Manag Rev 56: 149–151

Puaschunder JM (2016) On the emergence, current state, and future perspectives of Socially Responsible Investment (SRI). Consilience 16:38–63. http://www.jstor.org/stable/26188773

Qoyum A, Sakti MRP, Thaker HMT, AlHashfi RU (2022) Does the Islamic label indicate good environmental, social, and governance (ESG) performance? Evidence from sharia-compliant firms in Indonesia and Malaysia. Borsa Istanbul Rev 22(2):306–320. https://doi.org/10.1016/j.bir.2021.06.001

Rao A (2022) Deceptive claims using fake news advertising: the impact on consumers. J Market Res 59(3):534–554. https://doi-org.library.esc.edu/10.1177/00222437211039804

Reber B, Gold A, Gold S (2022) ESG disclosure and idiosyncratic risk in initial public offerings. J Bus Ethics 179:867–886. https://doi.org/10.1007/s10551-021-04847-8

Rowlands IH (1999) Corporate Potential [Review of Cannibals with Forks: The Triple Bottom Line of 21st Century Business, by J. Elkington]. Altern J 25(4):42–43. http://www.jstor.org/stable/45032233

Rubáček F, MacGregor Pelikánová R, MacGregor RK (2023) The sustainability of ESAs triumvirate for sustainability-related disclosures in the financial sector – all for one and one for all. Acta Universitatis Lodziensis. Folia Iuridica 105:123–143. https://doi.org/10.18778/0208-6069.105.08

Schüz M (2012) Sustainable corporate responsibility – the foundation of successful business in the new millennium. Central Eur Bus Rev 1(2):7–15

Siddiqui A, Altekar S, Kautish P et al (2023) Review of measurement of sustainable development goals: a comprehensive bibliometric and visualized analysis. Environ Sci Pollut Res 30:91761–91779. https://doi.org/10.1007/s11356-023-28887-x

Škapa S, Bočková N, Doubravský K, Dohnal M (2022) Fuzzy confrontations of models of ESG investing versus non-ESG investing based on artificial intelligence algorithms. J Sustain Finance Invest 13(1):763–775. https://doi.org/10.1080/20430795.2022.2030666

Skvarciany V, Lapinskaite I, Volskyte G (2021) Circular economy as assistance for sustainable development in OECD countries. Oeconomia Copernicana 12(1):11–34. https://doi.org/10.24136/oc.2021.001

Stepien S, Polcyn J (2019) Risk management in small family farms in Poland. Econ Sci Rural Dev 50:382–388. https://doi.org/10.22616/ESRD.2019.048

Strand R, Freeman RE, Hockerts K (2015) Corporate social responsibility and sustainability in Scandinavia: an overview. J Bus Ethics 127:1–15. https://doi.org/10.1007/s10551-014-2224-6

Taušer J, Artlová M, Žamberský P (2015) Czech exports and German GDP: a closer look. Prague Econ Pap 24(1):17–37. https://doi.org/10.18267/j.pep.498

Teng X, Wu KS, Kuo L, Chang BG (2023) Investigating the double-edged sword effect of environmental, social and governance practices on corporate risk-taking in the high-tech industry. Oeconomia Copernicana 14(2):511–549. https://doi.org/10.24136/oc.2023.014

Van Der Heijden HA (1999) Environmental movements, ecological modernisation and political opportunity structures. Environ Polit 8:199–221. https://doi.org/10.1080/09644019908414444

Van Tulder R, Van Mil E (2023) Principles of sustainable business. Frameworks for corporate action on the SDGs. Routledge, London and New York

White MA (2013) Sustainability: I know it when I see it. Ecol Econ 86:213–217. https://doi.org/10.1016/j.ecolecon.2012.12.020

Wójcik P (2016) How creating shared value differs from corporate social responsibility. J Manag Bus Adm. Central Europe 24:32–55. https://doi.org/10.7206/jmba.ce.2450-7814.168

Zatonatska T, Soboliev O, Zatonatskiy D, Dluhopolska T, Rutkowski M, Rak N (2024) A comprehensive analysis of the best practices in applying environmental, social, and governance criteria within the energy sector. Energies 17:2950. https://doi.org/10.3390/en17122950

Chapter 10
Conclusions

Ready or not, for better or for worse, sustainability is an integral part of our European way of life and perhaps the only key for the survival of the whole of humanity. Indeed, sustainability has an ideological dimension as well as a pragmatic dimension and this leads to a suggestion that sustainability concerns are embedded in our history and cultures, either directly or indirectly, actively or passively. Sustainability cannot be overlooked and underplayed and its importance questioned. At the same time, many questions and issues are linked to Sustainability, its understanding and regime. These questions come from different perspectives and angles and entail spheres studied by various branches and sub-branches of social science.

Certainly, sustainability is currently omnipresent and a discourse about it is not a novelty by any means. However, often such a discourse is reduced to a mere focus on only one aspect at the tip of the iceberg. Indeed, it is challenging to engage with a multidisciplinary study of the multitude of aspects of current sustainability and perhaps it is even more difficult to assess its spontaneous and very contextual evolution. However, without doing it, sustainability is perceived as a simplistic and obvious cliché linked to Western civilization and which has dark sides, such as a lack of pragmatism, ecological terrorism, Wester human rights imperialism, etc. Indeed, as much as sustainability is worshiped and showered by laudatory proclamations and declarations, so much is also under sharp challenges from many sides by many subjects. Sometimes this is done directly, sometimes indirectly, sometimes openly and sometimes secretly, sometimes with more legitimacy sometimes with less. Nevertheless, even the critics recognize that there is no doubt that sustainability is both an integral part of the current European way of life and also is appreciated by a significant part of Europeans as one of the keys, if not as the key, for the future. EU strategies, politics and law definitely believe in it and it also extends further global endeavours launched by UN in this arena. However, it would be very naïve and misleading to suggest that this is the unified European tenor. So what is the meaning and reality of sustainability and its regime in Europe? What is correct and what is wrong? How is it right and how is it wrong? Is sustainability a reality, a myth or a chimera in Europe? How do strategies, politics and law address it in Europe? Do we

R. MacGregor Pelikánová, *Sustainability in Europe*,
https://doi.org/10.1007/978-3-031-74689-5_10

have a pro-sustainability or anti-sustainability or no-sustainability regime in Europe? Is it legitimate, pragmatic, effective and efficient? These and many further questions emerge and typically one is picked up and answered in an isolated manner. This is logical, since this phenomenon is rather a bundle of many complex phenomena with a myriad of often controversial aspects lacking black-and-white unanimous answers. It would be foolish and extremely superficial to attempt, in one shot, to deal with all of that. Manifestly, a step by step systematic and academically robust approach is necessary within a multidisciplinary study entailing piecemeal instruments. To put it differently, in order to achieve a deep appreciation of sustainability in the current European setting and its current regime, it is necessary to resist the temptation to merely select the newest law, and perhaps as well policy, instruments and, based on their literate interpretation, extract the only and absolutely clear definition of sustainability and all rules about it with an easy interpretation and application. Even a cursory study of current sources and mere empirical observations shows that shortcuts do not work here at all and that an engagement in depth with the contextual multidisciplinary evolution is necessary. It is necessary to fully admit that sustainability is neither new nor developed by a plan nor fixed. Instead, sustainability was shaped by its spontaneous evolution closely linked to social and economic developments with a myriad of heterogenous, often unpredictable factors and strong personalities. It has dynamically changed its meaning, form and content, to become a rather complex phenomena definitely belonging to our European present and at the same time keeping both its (allegedly) obscure and continuously dynamically changing meaning and regime. The study of such a dynamic and ephemeral phenomena might appear as leading to a set of massive contradictions, but in reality there is a logic and sense to be extracted from the contextual and critical appreciation of its development by a set of branches of social sciences. In sum, we need to check the roots and understand the evolution, in particular various simultaneous trends, to complete our very first step on the long, challenging and desperately needed journey towards the real understanding of sustainability in Europe. Indeed, the past of sustainability in Europe is definitely not to be passed over, instead it is a *conditio sine qua non* and an inevitable pre-requirement for a genuine engagement with its current meaning and the setting and the related interpretation and application of its regime. This is the research aim which was projected in two mutually related questions: (i) what is the foundation, development and meaning of the concept of sustainability in Europe and (ii) what kind of regime and duties has it involved in Europe?

This monograph is addressing such an aim and these two questions in a systematic manner via ten chapters which are both related and sufficiently independent so that the sequential reading is recommendable but not necessary. The first chapter introduces the entire topic, offers definitions and does not hesitate to summarize the map and spectrum of sustainability and its regime by mentioning key legal documents and sister concepts. It is a rather robust chapter ambitiously already suggesting the pivotal importance of the spontaneous contextual evolution with roots going back thousands of years.

The second chapter presents the systematic manner and tools about how to address the aim and both questions and, in general, such a massive and complex phenomenon or, better to say, a set of phenomena. Often chapters about materials and methods are done in a mechanical and repetitive manner just referencing to one database and some established methods and formulas. Here, it is different. Since the term sustainability is an abstract, ephemeral and continuously evolving omnipresent expression, several branches of social sciences must be considered and, of course, each of them has its sources and methods. Therefore, each of these two questions is to be addressed by a set of various sources and a set of various methodology instruments, which are predominantly qualitative, multidisciplinary and multijurisdictional. The underlying drive and desire is to identify trends, tensions and their potential for synergetic reconciliation.

The third chapter organically builds on the second chapter and its methodology ramifications by focusing on the terminology and scientific background of sustainability. Sustainability is a term used differently in different eras in different jurisdictions by different branches of social science. The same applies even to its sister and supplementary concepts. Emphasis must be given to the often overlooked linguistics study, in particular the morphology, semantics and pragmatism of sustainability. Is the current sustainability more similar to, or different from, the original German *Nachhaltigkeit* and French *soutenabilité* or *durabilité*? This leads to the appreciation of the very historical and philosophical background and, in particular, the ontology from metaphysics (reality and existence of sustainability) and ethics from axiology (sustainability morally right). This implies the current meta-sustainability which had a rather simplistic, autocratic and pragmatically agricultural beginning.

The fourth chapter picks up exactly where the third chapter ended, i.e. it deals with the historical origins of sustainability. It covers the philosophical dimension of sustainability in antiquity with a focus on the virtue feature. The balancing and various types of justice by Aristotle, the Bible and its parables as well as historical and more modern academic writings are considered. The connection between sustainability in Europe and Christianity is undeniable as well as the progressive incorporation of a specific command to sustainability in politics and law in antiquity. Hammurabi's flooding management and Code, Roman farming and the Law of the Twelve Tables plus *Corpus Iuris Civilis* and the emerging network of monasteries and universities paved the way for sustainability even further.

The fifth chapter described this way, which was rather a set of various simultaneous journeys in different parts of Europe. Charlemagne can be perceived not only as the *Pater Europae*, but as well as one of the truly European sustainability founders and his work was continued by Capetian, Valois and Bourbon kings in France. Their pre-occupation with forest sustainability led to law statutes, policies and even the iconic quote by Colbert—*La France périra faute de bois*. In Northern Italy there flourished the study of the Roman law as well as textile production leading to other sets of sustainability concerns. In central Europe, sustainability began to be appreciated, and not only in relation to tangible natural resources, and this led to the addition of social, rural and even economic development aspects, both in general and in regard to special industries such as pond and fishery management. All of these teachings, in particular the French and English pre-occupation with diminishing forests and the Hansa business pragmatism, led to the Saxon cradle of modern European sustainability—the famous *Nachhaltigkeit.* Sustainability definitely changed from a local ad hoc agricultural concern of the leader into a regional systematic strategy extending into various fields and support for various industries and having its own methodology, see the endeavours by Carlowitz. Even more importantly, the original accountability turned into a responsibility and even a liability. However, a massive revolutionary challenge for sustainability came. The First Industrial Revolution rocked the entire establishment. The sustainability was pushed back, but did not disappear, as can be observed by works by Smith and Hegel. Later on, during the Second Industrial Revolution it managed to move to the . This development was radically impacted by Marx's vulgarizing Smith's teachings and looking for the revolution and abolition of the state. Interestingly, despite dramatic improvements in the social sphere, see Bismarck's welfare state legislation, in the environmental sphere, see the American National Parks and Forests legislation, and even in the economic sphere, see the general increase in the standard of living, the general suspicion about the capacity of the capitalist society to go for sustainability kept being questioned. A good study case is the example of the fashion industry. Well, much worse was ahead—the First World War and the Second World War, but thereafter came the UN with its sustainability crusade which continues until today and in which gladly and enthusiastically participates the EU.

The sixth chapter deals exactly with this, i.e. with the modern history of sustainability resting on three pillars (economic, environmental and social) and its milestones—the UN Charter, UDHR, Brundtland Report, Agenda 2030 with SDGs, etc. In addition, it points to the development of the shareholder theory, stakeholder theory and CSR with CSV. Sustainability is becoming a global multi-responsibilities command, if not directly a need. Therefore, the sixth chapter ends the chronological and rather general part, which is followed by a practical part entailing three chapters addressing particular aspects of modern sustainability.

The seventh chapter tackles the dynamic aspect of sustainability and its link to development, in particular the understanding, development and regime of sustainable development by the UN and EU. Arguably, this is about the contradiction implied by the linguistic setting of such a sustainability in motion. Well, again it is necessary to reveal roots and consider the history, and then move to a thorough analysis of the view of the UN and the EU. The famous Brundtland Report and the EU constitutional trio are pivotal, but as well secondary, sources of the EU law, the case law and even policy instruments, which all can help with the clarification. The interest and positive attitude of the European Commission to sustainability and sustainable development is obvious, but what about Europeans in general?

The eighth chapter reacts to this question and follows the discourse regarding the multi-stakeholder model for sustainability, in particular the call for private subjects to participate—CSR. Therefore, the eighth chapter organically builds on prior chapters and notes about CSR and engages in depth with CSR origins, visualization, stages, phases and versions, and ultimately as well appreciation.

The ninth chapter narrows the topic further and basically tackles the measurability of sustainability and CSR as needed by investors—ESG. The history of sustainability goes for millennia, the history of CSR for centuries and the history of ESG for decades. Namely, investors, as powerful stakeholders, have always made their decisions based on a set of factors, not all of them financial. However, a systematic consideration, and thus the implied need for appropriate information about sustainability and CSR, started perhaps in 1972 with the influential book by the Meadows, The limits to growth. After the initial period came the era of self-market regulation which finally was replaced by the current era of ESG standardization by the UN and by the EU.

In sum, regarding the first question, these nine chapters demonstrate the importance of the study of the roots and evolution and they reveal the transformation of sustainability from a rather narrow accountability command of leaders in a local setting into a strong multi-spectral and multi-sectorial supra-regional call for the responsibility of all subjects with a legal regime which progressively regulates its formal aspects (reporting), even partially its content aspects, and expands the pool of subjects to be induced to embrace such a newly emerging liability. Indeed, the exploration of the ancient roots and millennia long spontaneous evolution of the concept of sustainability and its regime, reveals how sustainability and our commitment to it is intimately linked to our civilization and history, to contextual events and to the eternal search of balance and justice. There were linear as well as cyclical phases with clear trends towards expanding the reach, the strictness of duties and also the pool of subjects. Therefore, a systematic knowledge of history, law and economics, as well as politics and policies, is indispensable in order to contextually figure out the meaning and the regime of sustainability in Europe. It allows to see the links and reasons, this offers suggestions, if not directly answers, to many current puzzling questions, especially related to issues of the (alleged) need of balancing, trading-off, compromising and missing pragmatism and feasibility. Yes, the concept of sustainability was always linked to duties and relied on the support by the social order, often transformed into the legal regime. However, sustainability was never about undermining the very existence. In sum, neither the foundations of sustainability nor sustainability during the course of the last millennia were contradicting reality. But, by being a constant in the background, and slowly making improvements in increasing fields of endeavour, it began to rise to the fore. It ultimately led to the interest in sustainable development and to ethical commands, turning into law various duties to respect, to avoid waste and to no longer indulge the temptation of immediate gratification at any price. Although originally concerns were given more to its presence and impacts in the agricultural local setting, while, over time, the industries' horizons progressively expanded as well as the geographic considerations, currently the driving engines are the UN and the EU with the European Commission. Both, the UN and EU, underline that sustainability and sustainable development have a clearly multi-generational global dimension, entailing all spheres of human activities and focusing on all three pillars. The drive from ethical to legal duties, from a facultative to a mandatory regime, from the national to the EU setting, is obvious. Similarly, obvious is the growing importance of the private law setting, the desire for a win-win CSV to be informally enforced even from the bottom, see the ESG and the EU law drive to expand the legal liability for providing information about sustainability and even to set compulsory sustainability targets and incorporate SDGs in the EU law.

The answer to the second question is basically about the appreciation of the roots and development addressed by the first question and about the inventory of sources from many jurisdictions. They indicate that it is no longer about which pillar is more important, but rather about how to reconcile the demands generated by each of these pillars. However, an effective and efficient reconciliation is appropriate only vis-à-vis genuine and legitimate demands. Foundations and values should serve as the compass and the prior regimes as guidelines, the past of Sustainability European Style is heterogenous and spontaneously evolving, taking advantage of our intellectual qualities reflected by creativity and responsibility. The current multi-stakeholder model, with calls for the rule of law and transparency, is fully justified, while the radical move from the soft law to hard law is, at least partially, questionable and the disregard of transparency and low awareness are not acceptable.

Sustainability in Europe should be more about motivating, inducing, facilitation and further developing the European inclination to advance (certain) values in an organic and patient manner, rather than jumping on "hot issues" and addressing one without knowing its past and the past of the paralleled and (allegedly) conflicting values. At the same time, considering the globalization and the condition of our planet, regarding certain aspects an action is needed and legal duty is needed. However, this leads to burning questions about what are the aspects needing mandatory hard regulation, how this will be enforced and most importantly how these duties should be set. The saga of the European Green Deal shows that, always, at least a partial consensus is needed. Such a consensus should be about the continuation of our civilization, and this consensus can be achieved only if Europeans understand the past, are ready to draw lessons from it, keep rejecting waste, recognize the importance of respect and can agree about fundamental values.

As a matter of fact, this is the very point and *raison d'être* of this monograph—to be a small glass fragment to help to complete the mosaic of understanding the roots and evolution of sustainability and to help to find a consensus about the regime. Hopefully, this monograph is a small part of the picture connecting the European past with the European future and reminding us about our shared values and virtues. The individual responsibility and creativity as well as the respect, free will, pragmatism and capacity to resist the temptation of immediate selfish self-gratification and other Christianity features and aspects paved our rather successful journey. Why leave it?! So let's appreciate it and, as appropriate, continue on with it. *Per ardua ad astra!*

The manufacturer's authorised representative in the EU is Springer Nature Customer Service Centre GmbH, Europaplatz 3, 69115 Heidelberg, Germany. If you have any concerns regarding our products, please contact ProductSafety@springernature.com

Printed and bound by CPI Group (UK) Ltd, Croydon, CR0 4YY
01/10/2026
02225909-0001